Threatened Birds of Uttar Pradesh

Threatened Birds of Uttar Pradesh

Asad R. Rahmani and Sanjay Kumar

with major contributions by
Neeraj Srivastav, Rajat Bhargava, and Noor I. Khan

Maps prepared by

Mohit Kalra and **Noor I. Khan**

Layout and design by

V. Gopi Naidu

Sponsored by

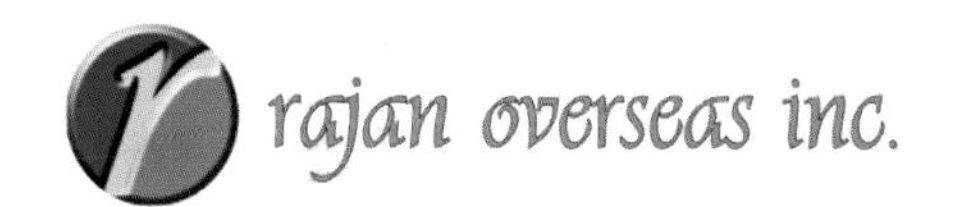

OXFORD

Oxford University Press, Walton Street, Oxford OX2 6DP
Oxford, New York,
Athens, Auckland, Bangkok,
Cape Town, Chennai, Dar-es-Salaam,
Delhi, Florence, Hong Kong, Istanbul, Karachi,
Kolkata, Kuala Lumpur, Madrid, Melbourne,
Mexico City, Mumbai, Nairobi, Paris,
Singapore, Taipei, Tokyo, Toronto,
and associated companies in
Berlin, Ibadan

Recommended citation: Rahmani, A.R., Kumar, S., Srivastav, N., Bhargava, R., and Khan, N.I. (2014) *Threatened Birds of Uttar Pradesh* Indian Bird Conservation Network, Bombay Natural History Society, Royal Society for the Protection of Birds, and BirdLife International. Oxford University Press. Pp. xiv + 226.

Text Editor: Gayatri Ugra

Layout and design: V. Gopi Naidu

Maps: Mohit Kalra and Noor I. Khan

IBCN, c/o BNHS, Hornbill House, Shaheed Bhagat Singh Road, Mumbai 400 001, India
Telephone: 0091-22-22821811, Fax: 0091-22-22837615
Email: ibabnhs@gmail.com, info@bnhs.org
Websites: <www.ibcn.in> <www.bnhs.org>

Bombay Natural History Society is registered in India under the Bombay Public Trust Act 1950: F244 (Bom) dated July 6, 1953

ISBN : 9780199455249

Proceeds from the sale of this book will go to the Indian Bird Conservation Network

Front Cover : Sarus Crane by Vandan Jhaveri
Back Cover : Black-necked Stork by K.S. Gopi Sundar

Available from
IBCN, c/o BNHS, Hombill House, Shaheed Bhagat Singh Road, Mumbai 400 001, India
Telephone: 0091-22-22821811, Fax: 0091-22-22837615
Email: ibabnhs@gmail.com, info@bnhs.org
Websites: <www.ibcn.in> <www.bnhs.org>

Processed by Trendz Phototypesetters. Email: gotrendz@gmail.com
Printed by Specific Assignments India Pvt. Ltd. Email: info@specificassignments.com

CONTENTS

CONTENTS (*contd.*)

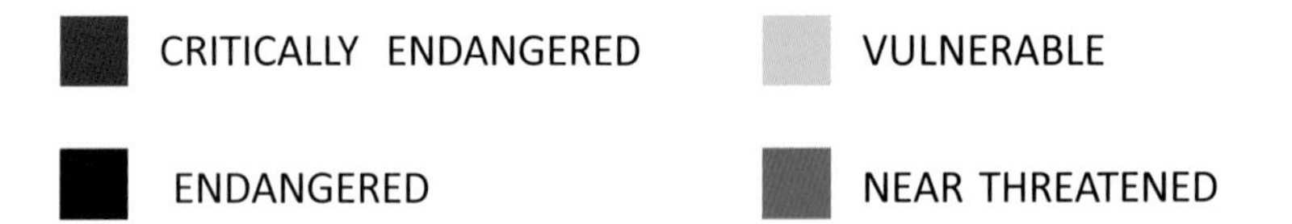

PREFACE

Uttar Pradesh is one of the largest states of India and has less than 7% forest cover. However, birdlife is very rich in the state mainly due to the presence of a large number of wetlands the Gangetic plains. Out of the 1,303 bird species reported from India, nearly 500 species e found in the state. Every year, BirdLife International, UK, brings out an updated list of globally reatened species of birds for IUCN. The 2013 list, on which this book is largely based, has 165 ecies of globally threatened and near threatened categories in India. Of these, about 50 are und in Uttar Pradesh. Three species, Great Indian Bustard, Pink-headed Duck, and Siberian ane are now extinct in the State. The Pink-headed Duck is extinct worldwide, and the Great dian Bustard is still surviving in other states in highly depleted numbers. The Siberian Crane, nigratory species, was last seen in UP in 1937, and in Keoladeo National Park, Rajasthan in e winter of 2001–2012.

We have made an attempt to describe all the globally threatened and near threatened ecies which are still found in the state. Although most of the species are found in other parts India, and many in Southeast Asia, UP is crucial to the survival of species such as the White-cked Vulture, Slender-billed Vulture, Swamp Francolin, Sarus Crane, Bengal Florican, Indian immer, Black-bellied Tern, Bristled Grassbird, Rufous-rumped Grassbird, and Yellow Weaver. e major species are described in detail, while marginal species which have limited historical d present records are briefly dealt with.

Uttar Pradesh has a good network of protected areas, particularly in the Terai. Dudhwa ger Reserve consists of Dudhwa National Park, Kishanpur Wildlife Sanctuary, and Katerniaghat LS. A new Tiger Reserve was declared in Pilibhit while this book was going to press. Out of e 466 Important Bird Areas (IBAs) in India, Uttar Pradesh has 25. Many IBAs are protected as nctuaries and national parks, but some are not legally protected. Most of the IBAs of Uttar adesh are important for threatened and near threatened species, particularly those species r which site-based conservation action is important. For many species that live in a larger ndscape, such as the White-backed Vulture and Sarus Crane, conservation initiatives have to taken at a different level.

Uttar Pradesh is known for its fabulous wetlands, sadly now mostly encroached and drained. one time, the state used to attract millions of waterfowl. Some wetlands still do, such as ndi, Lakh-Bahosi, Sur Sarovar, Samaspur, Sheikha Jheel, and Patna Bird Sanctuary, but most etlands are in dire straits. With proper restoration and sustainable harvest of resources, these etlands can play an important role in poverty reduction and biodiversity conservation.

We have given general recommendations in the Introduction and also specific commendations under each species. We believe that bird tourism, a growing business in the orld, should be promoted in the state and can be used as a conservation tool. With easy cessibility from Delhi, a very good rail and road network, and a list of nearly 500 bird species, can become an important destination for bird tourists.

We hope that our book will generate interest among decision makers, researchers, students, achers, and civil society at large. If the status of threatened birds improves, we will be satisfied at our attempt has not been in vain.

uthors

AKHILESH YADAV

LAL BAHADUR SHASTRI BHAWAN
LUCKNOW

CHIEF MINISTER
UTTAR PRADESH

May, 2014

MESSAGE

Uttar Pradesh is blessed with a large number of bird species, and some of the finest wetlands of India. Nearly 500 different types of birds are reported from the state. Unfortunately, nearly 50 species are under various types of threats to their survival, mainly due to human related activities. We need to find ways to protect the avian wealth of our state.

The present book *Threatened Birds of Uttar Pradesh* by Dr. Asad R. Rahmani, Mr. Sanjay Kumar, and others will create interest and also guide us on how to protect the birds. I congratulate the authors and the Bombay Natural History Society for bringing out this lovely book, with very good images of threatened birds. I hope a Hindi version will soon be brought out, so that the message of the book reaches the village level. We need more such books about the biodiversity of our state. I also hope the book will be a useful tool to the Forest Department for taking appropriate action for the protection of birds.

(Akhilesh yadav)

Jawed Usmani
I.A.S.

Ph. : (0522) 2621599, 2238212 (O)
: (0522) 2239283 (Fax)
: (0522) 2239461 (Res.)
Government of Uttar Pradesh
(Lal Bahadur Shastri Bhawan)
Lucknow – 226001
E-mail : csup@nic.in

MESSAGE

Public support for conservation of biodiversity is essential. This support will come with awareness, sustainable utilisation, and appreciation of the need to protect biodiversity. With millions of people taking up birdwatching as a hobby, birds are the most loved group of animals in the world. Uttar Pradesh is quite rich in bird diversity, with nearly 500 different species identified till date, and many more to be found. Unfortunately, some species are on the verge of extinction due to human activities. The UP Government is fully committed to protect the rich avian diversity of the state.

This book, which has been published by Bombay Natural History Society (BNHS) and Oxford University Press, with support from many national and international organizations, is being launched at an appropriate time when rapid industrialisation and socio-economic development have the potential to threaten avian sanctuaries in the state. The recommendations given in this book would help us in safeguarding UP's avian wealth.

Birdwatching is one of the fastest growing hobbies and professions in the world. With proper policy safeguards and administrative interventions, birdwatching can be used as a conservation tool. The state has 24 Protected Areas (PAs) maintained by the Forest Department and 25 Important Bird Areas (IBAs) identified by the BNHS and BirdLife International. These PAs and IBAs can attract thousands of visitors and birdwatchers, providing livelihood and employment to the local populace.

I hope that the book will be used by decision makers, birdwatchers, conservationists, and the general public to protect the birdlife of Uttar Pradesh, particularly those bird species that face the danger of extinction.

(Jawed Usmani)

MESSAGE

I am very happy that the Bombay Natural History Society (BNHS), particularly Dr. Asad R. Rahmani, Director, BNHS along with his colleagues, is bringing out a book on threatened birds of Uttar Pradesh. The birdlife of Uttar Pradesh is amazingly diverse, with nearly 500 species reported till now. With more research and monitoring, I am sure the list will go up.

Uttar Pradesh has one national park and 25 sanctuaries, of which 12 have been established specifically to protect birdlife. Notable among them are Sandi, Lakh-Bahosi, Patna Jheel, Sur Sarovar, Samaspur, Surha Taal, and Nawabganj. Some other important wetlands that serve as habitat for birds are also under conservation cover.

BNHS and BirdLife International have identified 466 Important Bird Areas in India and the state contributes 25 IBAs to this list. Based on birdlife, many existing PAs were identified as IBAs. It is rather unfortunate that nearly 50 bird species of Uttar Pradesh are listed as globally threatened and near threatened by BirdLife and IUCN. The state is fully committed to protect them. The network of PAs and IBAs can secure the future of many such species, but for species that live in larger landscapes, such as our State Bird, Sarus Crane, we need to have a landscape level approach towards conservation. The state hosts nearly 60% of the Sarus Crane population of India. To protect Sarus and its wetland habitat, Uttar Pradesh has established Sarus Protection Society which is doing yeoman service. The Sarus Protection Society has full support of the Government of Uttar Pradesh.

I am sure this book will go a long way in focusing attention on the urgent need to ensure that our threatened birds are conserved for posterity.

Dr. Rupak De
Principal Chief Conservator of Forests and
Chief Wildlife Warden, Uttar Pradesh

FOREWORD

The Mission statement of BNHS-India says "Conservation of nature, primarily biological diversity, through action based on Research, Education and Public Awareness". These fifteen words summarise the work that BNHS has been doing for the last 130 years. BNHS's scientific journal, magazines, and newsletters, and now the BNHS website, help in generating public awareness and interest in wildlife conservation. BNHS has a strong scientific base on which an edifice of conservation action and public awareness stands.

BNHS also believes in cooperation and collaboration, both national and international. The Important Bird Areas Programme is a fine example of this collaboration. Under this Programme, in 2012 we brought out a landmark publication by our Director, Dr. Asad R. Rahmani, titled *Threatened Birds of India: Their Conservation Requirements*. The book was sold out within two years. As this book was a large 861-page publication, it was soon felt that smaller state-wise books would be necessary, as conservation action takes place mostly at the state level. During the last two years, four state-wise books on threatened birds have been brought out by us. *Threatened Birds of Uttar Pradesh* is the fifth in this series.

The birdlife of India is very rich with nearly 1,220 species described from our territorial limits. With more birdwatchers and better equipment, more species are added every year. Unfortunately, the list of threatened and near threatened species is also increasing, and according to BirdLife International, of which BNHS is a Partner, 165 bird species found in India qualify for the 2013 IUCN Red List. Of these, nearly 50 species are found in Uttar Pradesh. This book describes each species in brief and gives recommendations for their protection. General recommendations for protection all bird species are also given in the Introduction.

I hope the book will be useful to the Forest Department of Uttar Pradesh, scientists, conservationists, and birdwatchers. The purpose of this book will be served if the declining trend of bird species is reversed. We can then proudly say that we have achieved our Mission.

Homi Khusrokhan
President, BNHS-India

ACKNOWLEDGEMENTS

The authors wish to thank Rajan Overseas Inc. for providing financial support for the publication of this book.

We thank the UP Forest Department for its support. The first author who has been interacting with the UP Forest Department for more than 35 years would like to thank all the former Chief Wildlife Wardens, starting from Mr. V.B. Singh to Mr. B.K. Patnaik. Particular reference has to be made to the following CWLWs for their support: Mr. G.P. Singh, Mr. D.N. Lohani, Mr. M.D. Upadhayay, Mr. V.B. Singh, Mr. C.B. Singh, Mr. S.S. Srivastava, Mr. R.D. Gupta, Mr. R.P. Sharma, Mr. R.C. Bhadauria, Mr. M.C. Ghildiyal, Mr. Ashok Singh, Dr. Ram Lakhan Singh, Mr. Anil Berry, Mr. D.N. Bhatt, Mr. Mohammad Ahsan, Mr. D.N.S. Suman and Mr. B.K. Patnaik.

We are grateful to Dr. Rupak De, present Chief Wildlife Warden and Principal Chief Conservator of Forests (Wildlife), and all his staff for support to our work. Our gratitude to Mr. V.N. Garg, Principal Secretary (Forest) for his encouragement. Special thanks to to Ms Eva Sharma, Ms Renu Singh, and Ms Pratibha Singh.

We are thankful to the officers of the UP Forest Department, especially Mr. M.P. Singh, Mr. K.K. Singh, Mr. Neeraj Kumar, Mr. Maneesh Mittal, Mr. P.P. Singh, Mr. Sanjay Srivastava, Mr. A.K. Singh, Mr. Shailesh Prasad, Mr. Sanjay Pathak, Mr Sujoy Bannerjee, Mr. N.K. Janoo, Mr. R.S. Mishra, Dr. Dhananjai Mohan, and Mr. R.C. Jha. We also extend our gratitude to the following Forest Officers: Mr. Suresh Kumar, Mr. T.R. Dohre, Mr. S.P. Singh, Mr. A.K. Tripathi, Mr. Manoj Kumar Shukla, Mr. Irfan Khan, Mr. Chandreshwar Singh, Mr. R.B. Uttam, Mr. Amit Katiyar, Mr. Vivek Gupta, Mr. R.B. Gupta, Mr. R.K. Gupta, Mr. Atul Kumar Aggarwal, Mr. U.S. Dhore, Mr. V.N. Thakur, Mr. Karan Singh, Mr. Ram Harak, Mr. Sunil Kumar, Mr. Ishaq, Mr. D.P. Rai, Mr. Narsingh, Mr. V.K. Sachan, Mr. Shivaji Rai, Mr. Saryu Prasad, Mr. D.K. Mishra, Mr. R.K. Sachan, Mr. U.P. Singh, Mr. Mukesh Kumar, Mr. R.S. Mishra, Mr. Amresh Chandra, Mr. Ajay S. Pandey, Mr. Prakash Shukla, Mr. Ashish Tiwari, Mr. A.K. Shrivastava, Shri V.K. Pathak, Mr. Shrinath Yadav, and Mr. Ramesh.

We are extremely grateful to Mr. Rigzin Samphel, Special Secretary to the Chief Minister of Uttar Pradesh.

We acknowledge the help of Ms Niharika Singh, Dr. Jaswant Singh Kaler, Dr. Kaajal Dasgupta, Dr. Amit Pathak, Mrs. Rita Seth, Mr. Arvind Nath Seth, Mr. Siddharth Sharma, Dr. Amita Khurana, Major Prithvi Rathore, Mr. Amit Puri, Mr. Vijay Kutty, Mr. P.N. Singh, Dr. Pradeep Panwar, Mr. Lalit Kumar Verma, Mr. Prakash Shukla, Mohammed Aklaq, Leeladhar (Sonu), Mr. Naseem, Mr. Dubeer Hasan, Dr. Harish Gularia, Mr. Raja Mandal, Mr. Jatinder Pandy, and Mr. Sunil Jaiswal.

We would like to thank Mr. Samir Sinha, Dr. S.K. Niraj and Mr. Abhinav Srihan.

In BNHS, we want to thank Dr. Gayatri Ugra for editing the text, Mr. Gopi Naidu for layout and design, and Mr. Mohit Kalra for preparing the maps. Our gratitude to Mr. Homi Khusrokhan, President, BNHS, Mrs. Sumaira Abdulali, Hon. Secretary, Mr. E.A. Kshirsagar, Hon. Treasurer, and all the members of the Governing Council for their support.

Among our colleagues at BNHS, we would like to thank Mr. Sachin Kulkarni, Dr. Raju Kasambe, Mr. Abhijit Malekar, Mr. Atul Sathe, Dr. Deepak Apte, Dr. V. Shubhalaxmi, Ms Varsha Chalke, Ms Vibhuti Dedhia, Ms Divya Varier, Mr. Divyesh Parikh, Mr. Isaac Kehimkar, Mr. J.P.K. Menon, Mr. M.G. Mathews, Mr. Mrugank Prabhu, Ms Neha Sinha, Ms Nirmala Barure, Ms Nisha Shilaj, Ms Parveen Shaikh, Mr. Rahul Khot, Mr. Sameer Bajaru, Mr. Santosh Mhapsekar, Mr. Siddhesh Surve, Mr. Asif Khan, Mr. Vandan Jhaveri, Ms Sonali P. Vadhavkar, and Dr. Swapna Prabhu.

We would like to thank the following people who contributed their valuable information about threatened bird species found in Uttar Pradesh which has been incorporated and duly acknowledged in the book:

Mr. Abrar Ahmed, Mr. Gobind Sagar Bhardwaj, Mr. Nikhil Bhopale, Mr. Gurmeet Singh, Mr. Khembahadur, Mr. Rakesh Kumar Prajapati, Mr. Raja Mandal, Mr. P.D. Mishra, Mr. Chandan Prateek, Mr. Amit Mishra, Mr. Satpal Singh, Dr. S.K. Srivastav, Mr. B.N. Singh, Mr. Abu Arshad Khan, Mr. Neeraj Mishra, Mr. Rishi Bajpayee, Mr. RBL. Uttam, Dr. V.P. Singh, Mr. Fazlur Rahman, Mr. Suresh Choudhury, Mr. B.C. Choudhury, Mr. K.S. Gopi Sundar, Dr. Satish Kumar, Mr. Nikhil Shinde, Mr. Mohammad Bilal, Mr. Bridesh Kumar, Mr. Vipin Agarwal, Mr. Satish Jain, Mr. Ankit and Ms Shruti Bhargava.

We wish to thank the following persons for contributing their excellent photographs for the making of this book: Mr. Dhritiman Mukherjee, Mr. Leeladhar (Sonu), Mr. Satish Kumar, Mr. Otto Pfister, Mr. Mohan Lal Meena, Mr. Bhasmang Mehta, Mr. Gobind Sagar Bhardwaj, Mr. Ramki Srinivasan, Mr. Saleel Tambe, Mr. Shashank Dalvi, Mr. Shrikant Ranade, Mr. Shyam Ghate, Mr. V. Gopi Naidu, Mr. Vinayak Yardi, Dr. Dharmendra Khandal, Mr. Satyendra Kumar Tiwari, Mr. Ashok Choudhary, Ms Ananya Mukherjee, Mr. Anand Arya, Mr. Rishad Naoroji, Mr. Niranjan Sant, Mr. Veer Vaibhav Mishra, Mr. Simon van Der Meulen, Mr. Nikhil Devasar, Mr. P.M. Lad, Dr. K.S. Gopi Sundar, Mr. Sunil Singhal, and Mr. Yogendra Shah.

We would like to thank BirdLife International and Royal Society for the Protection of Birds (RSPB), both based in UK, for their unstinted support to this book. In BirdLife International, the first name that comes to mind is that of Dr. Nigel Collar whose contribution to the study of threatened birds of the world is well-known. We thank Dr. Marco Lambertini, CEO of BirdLife International, Dr. Richard Grimmett, Dr. Mike Crosby, Dr. Stuart Butcher, Dr. A.J. Stattersfield, Dr. Richard Thomas, and Mr. Joe Taylor. In RSPB, we would like to acknowledge the support of Dr. Mike Clarke, CEO, Dr. Tim Stowe, Mr. Chris Bowden, Ms Cristi Nozawa, and Mr. Ian Barber.

■ ■ ■

DHRITIMAN MUKHERJEE

Bengal Florican: Pride of Uttar Pradesh

INTRODUCTION

The state of Uttar Pradesh (23° 52' - 30° 24'; 77° 5' - 84° 38' E) spreads over 2,40,972 sq. km, and comprises 7.33% of the country's total landmass. The state is surrounded by Uttarakhand and Nepal to the north, Haryana and Delhi to the west, Rajasthan to the south-west, Madhya Pradesh to the south and south-west, and Bihar to the east. It was created on April 1, 1937 as the United Provinces, and was renamed Uttar Pradesh in 1950. The state has a very ancient and interesting history. Many of the great sages of Vedic times including Bharadwaja, Yajnavalkya, Vashishta, Vishvamitra, and Valmiki lived in Uttar Pradesh. Many sacred books were also composed here. *Varaha Purana*, for example, is associated with Mathura. The sacred place of Naimisharanya in Sitapur district is where the sage Maharshi Vyas edited the Vedas and also where the Puranas were composed. Two great epics of India, the *Ramayana* and the *Mahabharata* have references to places in Uttar Pradesh. The *Mahabharata* tells the story of a royal family at Hastinapur, an ancient city located north-west of the state.

Uttar Pradesh is one of the most populous states of India. According to a report, the human population of the state was estimated at 166.1 million in the year 2001 and increased by 25.8% during the last decade. Its population density increased rapidly, from 300 persons per sq. km in 1971 to 689 per sq. km in 2001 and 828 per sq. km in 2011.

Dudhwa is the only National Park in Uttar Pradesh. It is famous for its majestic Sal forest

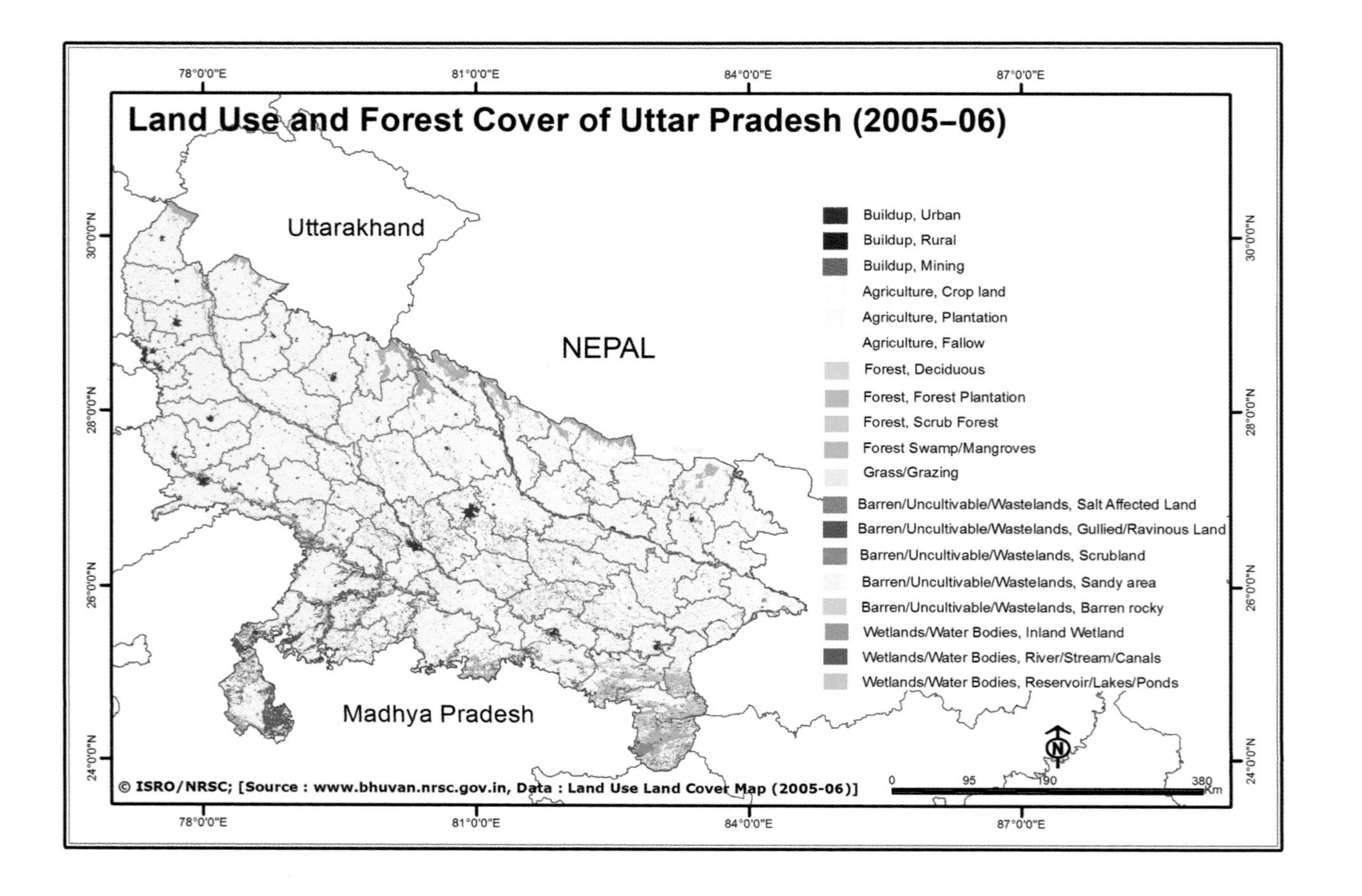

Land Use and Forest Cover of Uttar Pradesh (2005–06)
Uttarakhand
NEPAL
Madhya Pradesh
Buildup, Urban
Buildup, Rural
Buildup, Mining
Agriculture, Crop land
Agriculture, Plantation
Agriculture, Fallow
Forest, Deciduous
Forest, Forest Plantation
Forest, Scrub Forest
Forest Swamp/Mangroves
Grass/Grazing
Barren/Uncultivable/Wastelands, Salt Affected Land
Barren/Uncultivable/Wastelands, Gullied/Ravinous Land
Barren/Uncultivable/Wastelands, Scrubland
Barren/Uncultivable/Wastelands, Sandy area
Barren/Uncultivable/Wastelands, Barren rocky
Wetlands/Water Bodies, Inland Wetland
Wetlands/Water Bodies, River/Stream/Canals
Wetlands/Water Bodies, Reservoir/Lakes/Ponds
N
0
95
190
380
Km
78°0'0"E
81°0'0"E
84°0'0"E
87°0'0"E
30°0'0"N
28°0'0"N
26°0'0"N
24°0'0"N
© ISRO/NRSC; [Source : www.bhuvan.nrsc.gov.in, Data : Land Use Land Cover Map (2005-06)]

Uttar Pradesh is the largest producer of food grains and oilseeds in the country. The state leads in the production of wheat, maize, barley, grain, sugar cane and potatoes. Cities like Agra (famous for the Taj Mahal, Agra Fort, and nearby Fatehpur Sikri), Allahabad (confluence of the sacred rivers Ganga, Yamuna, and Saraswati, where Kumbha, the world's largest religious fair is held), Lucknow (seat of the Nawabs of Awadh, historical monuments), and Varanasi (Hindu pilgrimage and sacred places) have a strong historical and religious presence and international importance. Dudhwa National Park is the only national park of the state. It is also recognised as an Important Bird Area.

Geographical Profile

Uttar Pradesh has a tropical climate with a wide temperature fluctuation from less than 2 °C to 48 °C. There are three main seasons: summer from March to mid-June; the rainy season from mid-June to September; and winter from October to February. There is a great variation in rainfall. The *bhabhar* area has rainfall from 1,300 to 1,900 mm, whereas in the *terai* area it varies from 1,200 to 2,500 mm. In the Gangetic plains, the rainfall varies from 600 to 1200 mm.

Vegetation

Regarding their legal status, Reserved Forests constitute 65.9%, Protected Forests 14.4%, and Unclassed Forests 19.7% (Ministry of Environment and Forests 2001). There are three forest types, namely Tropical Moist Deciduous, Tropical Dry Deciduous, and Tropical Thorn. Sal is an important forest formation of the state. Forests are distributed largely in the northern and partly in the southern

ASAD R. RAHMANI

The grassland of the Terai, now much depleted, harbours many globally threatened animals

The Bengal Florican (female seen above), entirely depends on the grassland for foraging, display, and breeding

parts of the state. The central part is devoid of forest vegetation as it is mainly under agriculture. A forest cover increase was recorded by the Forest Survey of India report of 1999 in the districts of Hardoi, Lakhimpur-Kheri, and Saharanpur, because plantation was undertaken four to five years earlier and also due to effective protection measures. A decrease in forest cover was observed in the districts of Banda, Jhansi, Mirzapur, and Sonbhadra, which was largely on account of biotic pressures. According to the Forest Survey of India report of 2001, the recorded forest area in the state is 16,826,000 ha, about 2.2% of India's forest and 7% of the state's geographical area.

Prime Eco Zones

Uttar Pradesh has three main physiographic regions, namely the submontane region lying between the Himalaya and the plains, the vast alluvial Gangetic plains, and the southern hills and plateau. The state is fed by five major rivers: Ganga, Yamuna, Ramganga, Gomti, and Ghaghra. More than one-fourth of Uttar Pradesh lies within the Gangetic plains which consist of alluvial deposits brought down from the Himalaya by the Ganga, Yamuna, and their tributaries. All the rivers except the Gomti and Chambal emerge from the Himalaya. The southern hills form part of the Vindhya ranges whose elevation rarely exceeds 300 m.

Although grass is found in open wooded areas, it grows best in open sunny areas, called *phanta* and *chander* in the Terai

The Terai

The Terai ecosystem is one of the most diversified, but threatened ecosystems in the country. It supports many globally threatened bird species such as the Bengal Florican *Houbaropsis bengalensis*, Swamp Francolin *Francolinus gularis*, White-throated Bushchat *Saxicola insignis,* Bristled Grassbird *Chaetornis striata,* and an occasional Lesser Florican *Sypheotides indica*. The region is a vast flat alluvial plain lying between the Himalayan foothills and the Gangetic plains. It forms an integral part of the Terai-Bhabhar biogeographic subdivision of the upper Gangetic biotic province and the Gangetic Plains biogeographic zone (Rodgers & Panwar 1988)

Terai forest is mainly moist deciduous, and dominated by Sal *Shorea robusta*. A significant attribute of the Sal forest ecosystem is the interspersed swamp, wet tall grasslands, dry grasslands or *phanta* sparsely dominated by Kans grass *Saccharum spontaneum*, Blady grass *Imperata cylindrica*, and other specific vegetation. Broadly, two types of tall grasslands – the upland grassland and lowland grassland – have been recognized in the Terai region. The former occurs on drier or well-drained soils while the latter type occurs in low lying water-logged sites or areas inundated during the monsoon and subsequent months.

The grasses in upland grasslands usually attain a height up to 2 m while grasses in lowland grasslands are even 6 m tall. The resulting complex woodland-grassland-wetland ecosystem harbours a vast variety of floral and faunal life. The present extent of grassland forms an integral part of the forest land controlled by the state forest department.

ASAD R. RAHMANI

The vast Gangetic plains have numerous rain-fed wetlands which harbour a multitude of waterbirds (below). Some of the wetlands have been declared as waterbird sanctuaries, such as Sandi Bird Sanctuary shown above

RUPAK DE

Gangetic Plain

The most important area for the economy of the state is the Gangetic plain which stretches across the entire length of the state from west to east. The entire alluvial plain can be divided into three sub-regions. The first in the eastern tract consists of 14 districts which are subject to periodical floods and droughts and have been classified as scarcity areas. These districts have the highest density of population and the lowest per capita land. The other two regions, the central and the western, are comparatively better with a well-developed irrigation system. They suffer from water logging and large-scale *usar* (so-called wasteland) tracts. The Gangetic plain is watered by the Ganga, Yamuna, Ghaghra, Rapti, and Gandak. The major tributaries draining into these rivers are the Ramganga, Gomti, Hindon, Chambal, Saryu, Sai, Kosi, Betwa, Belan, Dhasan, Tons, and Son. Besides the rivers, the state is dotted with numerous waterbodies: lakes, ponds, reservoirs, dams, and canals. Twenty Important Bird Areas have been recognised in the Gangetic Plain. The whole plain is alluvial and very fertile. The chief crops cultivated here are rice, wheat, pearl millet, gram, and barley. Sugar cane is the chief cash crop of the region.

Southern hills and Plateau

The southern fringe of the Gangetic plain, from west to east, is demarcated by the Vindhya hills and plateau. It comprises districts Jhansi, Jalaun, Banda, and Hamirpur in Bundelkhand area, Meja and Karchhana *tehsils* of Allahabad district, and the whole of Mirzapur district south of the Ganga and Chakia *tehsil* of Chandauli district. The terrain is stony with low hills. The Betwa and Ken rivers join the Yamuna from the south-west in this region. It has four distinct kinds of soil, two of which are agriculturally difficult to manage. Rainfall is scanty and erratic, and water resources are scarce. Dry farming is practiced on a large scale.

NEERAJ SRIVASTAV

Large wading birds such as the Painted Stork depend on jheels, village ponds, and even roadside ditches for foraging

In 2004, BNHS and BirdLife International recognized 466 IBAs in India (Islam & Rahmani 2004), out of which 25 are found in Uttar Pradesh. During the last ten years, many more sites have come to our notice that fulfill one or more IBA criteria. Description of such sites will be given in the updated version of the book *Important Bird Areas in India* which is currently under preparation.

Birdlife of Uttar Pradesh

The birdlife of Uttar Pradesh is rich and varied. More than 500 species are found, including some extremely rare ones. BirdLife International (2013) has listed 16 Critically Endangered species in India, of which five are found in Uttar Pradesh: Oriental White-backed Vulture *Gyps bengalensis,* Long-billed Vulture *G. indicus*, Slender-billed Vulture *G. tenuirostris*, Red-headed Vulture *Sarcogyps calvus*, and Bengal Florican *Houbaropsis bengalensis*. Similarly, out of 15 Endangered bird species of India (Rahmani 2012), three are reported from Uttar Pradesh, of which Egyptian Vulture *Neophron pernopterus* has a good population in the state, while the other two, Lesser Florican *Sypheotides indica* and Baer's Pochard *Aythya baeri* have a few records. There is one record of the White-headed

SANJAY KUMAR

Four Critically Endangered vulture species are found in Uttar Pradesh. Twenty-five years ago, millions of White-backed and Long-billed Vultures used to be present in the state

Duck *Oxyura leucocephala* from Aama Khera, a wetland in Aligarh district (Z.A. Islam *pers. comm* 2010). The Greater Adjutant *Leptoptilos dubius* has not been recorded in the past many decades (Rahmani *et al*. 1990) but could occasionally occur in eastern Uttar Pradesh.

Out of the 53 Vulnerable species in India, 12 are reported from Uttar Pradesh, and for the following eight species, this state is very important for survival: Swamp Francolin *Francolinus gularis*, Sarus Crane *Grus antigone*, Pallas's Fish-eagle *Haliaeetus leucoryphus*, Indian Skimmer *Rynchops albicollis*, Great Slaty Woodpecker *Mulleripicus pulverulentus*, Bristled Grassbird *Chaetornis striata*,

SANJAY KUMAR

National Chambal Sanctuary which extends through Uttar Pradesh, Rajasthan, and Madhya Pradesh, is perhaps the best area in India to see Indian Skimmer

NEERAJ SRIVASTAV

Uttar Pradesh has many nesting colonies of waterbirds, the most famous being found in Sur Sarovar Sanctuary near Agra

DHRITIMAN MUKHERJEE

The Black-bellied Tern breeds on islands in large rivers such as Ganga, Yamuna, Chambal, and Ghaghra. Proper surveys are required to know its breeding distribution in Uttar Pradesh

RAJAT BHARGAVA

The Black-necked Stork is a popular display bird in zoos. With proper care it can breed in captivity

White-throated or Hodgson's Bushchat *Saxicola insignis*, and Yellow Weaver *Ploceus megarhynchus*.

BirdLife International (2013) has listed 75 Near Threatened bird species in India, 19 of which occur in Uttar Pradesh. For two species, the Black-necked Stork *Ephippiorhynchus asiaticus* and the Rufous-rumped Grassbird *Graminicola bengalensis* (earlier known as Large Grass-Warbler) wetlands and tall grasslands of Uttar Pradesh are very important for their survival. Earlier, Rahmani (1989), and recently Sundar & Kaur (2001) have shown that the wetlands of Uttar Pradesh are the major strongholds of the Black-necked Stork. It is found in 16 IBAs of Uttar Pradesh.

A species that needs special attention is the Hodgson's Bushchat *Saxicola insignis*. It is also known as the White-throated Bushchat or Hodgson's

Stonechat. It has a highly localised breeding range in the mountains of Mongolia where it is difficult to study. Its winter range is the northern Gangetic plains and the *duar*s of northern India, and the *terai* of Nepal and Uttar Pradesh. From the comparatively little information available, it is probably the scarcest species in its genus (Urquhart 2002). In northern India, it has been historically reported from Ambala in the west to northern Bengal in the east (Ali & Ripley 1987). It is found in heavy grassland, reeds, and tamarisk along riverbeds and cane fields. Earlier it was recorded in Kanpur, Gonda, Faizabad, Basti, and Gorakhpur (BirdLife International 2001, Urquhart 2002) but there is a recent record only from Corbett Tiger Reserve (Manoj Sharma *pers. comm.* 2010), now in Uttarakhand. Javed & Rahmani (1998) did not record it in Dudhwa. However, due to the development of tall grasslands and marshes on seepages of the vast canal systems of the state, and extant tall grasslands along major rivers, this species is likely to be present in many areas. Considering the paucity of reliable birdwatchers in Uttar Pradesh, it is most likely that the bird is found in many more areas than we know at present.

GLOBALLY THREATENED AND NEAR THREATENED BIRDS FOR WHICH UTTAR PRADESH IS VERY IMPORTANT

CRITICALLY ENDANGERED	
Slender-billed Vulture	*Gyps tenuirostris*
White-backed Vulture	*Gyps bengalensis*
Bengal Florican	*Houbaropsis bengalensis*
ENDANGERED	
Black-bellied Tern	*Sterna acuticauda*
VULNERABLE	
Pallas's Fish-eagle	*Haliaeetus leucoryphus*
Swamp Francolin	*Francolinus gularis*
Sarus Crane	*Grus antigone*
Indian Skimmer	*Rynchops albicollis*
Bristled Grassbird	*Chaetornis striata*
Yellow Weaver or Finn's Baya	*Ploceus megarhynchus*
NEAR THREATENED	
Black-necked Stork	*Ephippiorhynchus asiaticus*
Indian River Tern	*Sterna aurantia*
Rufous-rumped Grassbird	*Graminicola bengalensis*

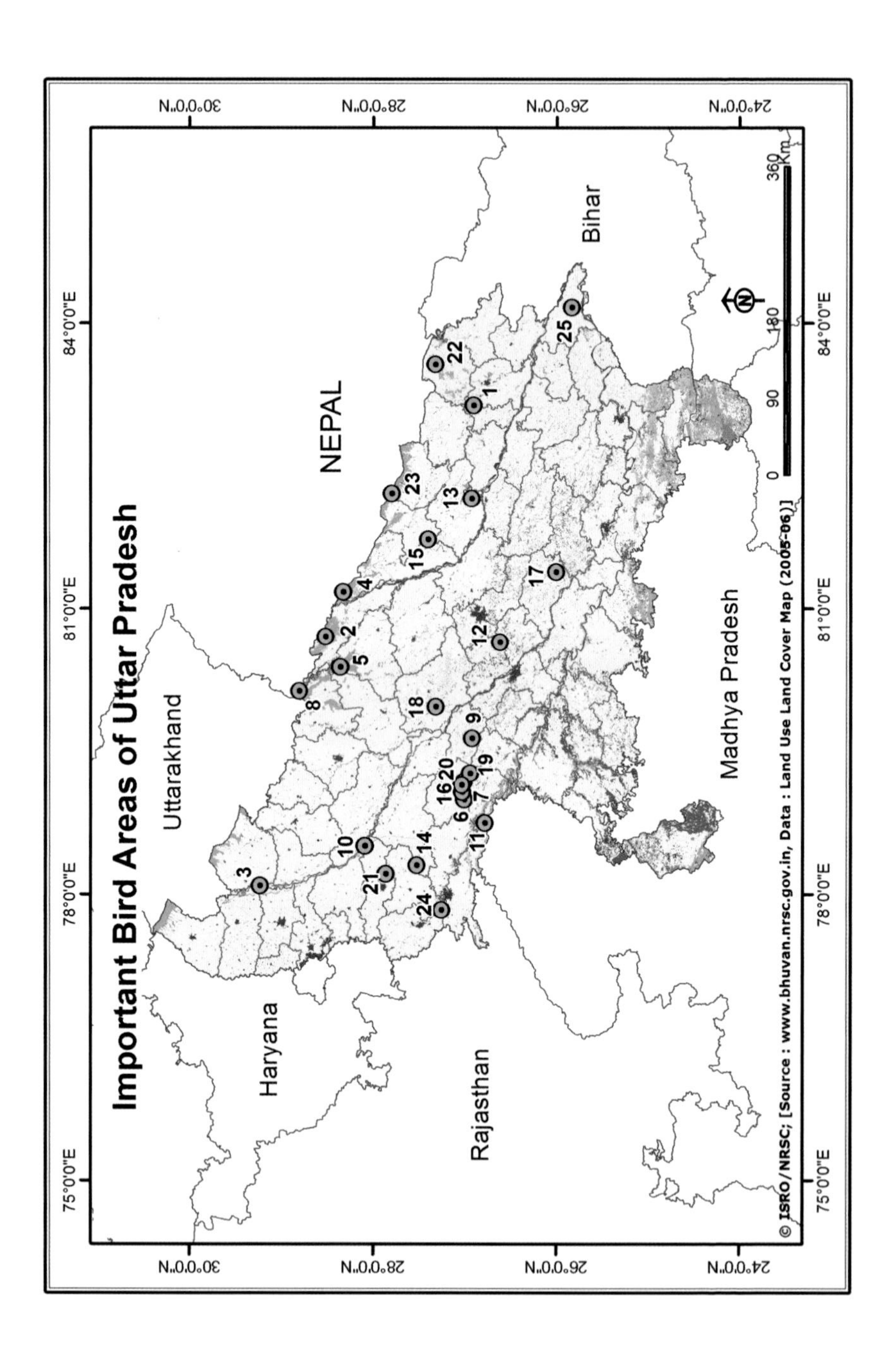

Important Bird Areas of Uttar Pradesh
NEPAL
Uttarakhand
Haryana
Rajasthan
Madhya Pradesh
Bihar
1
2
3
4
5
6
7
8
9
10
11
12
13
14
15
16
17
18
19
20
21
22
23
24
25
0 90 180 360 Km
30°0'0"N
28°0'0"N
26°0'0"N
24°0'0"N
75°0'0"E
78°0'0"E
81°0'0"E
84°0'0"E
© ISRO/NRSC; [Source : www.bhuvan.nrsc.gov.in, Data : Land Use Land Cover Map (2005-06)]

Important Bird Areas of Uttar Pradesh

The IBA Programme of BirdLife International aims to identify, monitor, and protect a global network of Important Bird Areas (IBAs) for the conservation of the world's birds and other biodiversity. BirdLife Partners take responsibility for the IBA Programme nationally, with the BirdLife Secretariat taking the lead on international aspects and in some priority non-Partner countries. As of 2013, more than 12,000 sites in some 200 countries and territories have been identified as Important Bird Areas.

Important Bird Areas are sites of international significance for the conservation of birds and their habitats at the global, regional, and sub-regional level. The selection of IBAs is a particularly effective way of identifying conservation priorities. IBAs are key sites for conservation, small enough to be conserved in their entirety and often already part of a protected area network (BirdLife International undated).

A site is recognised as an IBA only if it meets certain criteria based on the occurrence of key bird species that are vulnerable to global extinction or whose populations are otherwise irreplaceable. The IBA programme aims to fill the gaps in knowledge and further understanding about the conservation of these sites and their avifaunal diversity.

An IBA must be amenable to conservation action and management. The IBA criteria which are applicable globally are as follows:

A1: Sites holding globally threatened bird species of global conservation concern.

A2: Sites having restricted range bird species, i.e., bird species with a historic breeding range up to 50,000 sq. km in the world.

A3: Sites having biome restricted bird species, i.e., bird species representing distinct habitat types.

A4: Sites having large congregations of birds.

What is the significance of IBAs?

For conservation

- IBAs help identify priority sites for conservation action.
- IBAs provide the framework to monitor and manage sites of global conservation significance.
- IBAs provide decision makers with high quality information needed to formulate national conservation strategy and implement international agreements.
- IBA programme helps develop national and local capacity for biodiversity conservation.

For communities

- IBAs help meet daily subsistence needs of communities for food, fuel, fodder, and other natural resources.
- IBAs are a source of livelihood for many communities who harvest minor forest produce for sale in local markets.
- IBAs are a part of distinct indigenous cultures and a repository of traditional knowledge resources.

For climate change

- IBAs play an important biological role as carbon sinks, thereby reducing the amount of CO_2 in the atmosphere.
- IBAs help mitigate the impact of extreme weather events such as drought, flash floods, and cyclones by acting as a buffer for human habitations.
- IBAs help climate change affected communities cope by providing water, food, and building material for temporary shelters.

IBAs of Uttar Pradesh (Based on Islam & Rahmani 2004)		
IBA site codes	**IBA site names**	**IBA criteria**
IN-UP-01	Bakhira Wildlife Sanctuary	A1, A4iii
IN-UP-02	Dudhwa National Park	A1, A3
IN-UP-03	Hastinapur Wildlife Sanctuary	A1
IN-UP-04	Katerniaghat Wildlife Sanctuary and Girijapur Barrage	A1
IN-UP-05	Kishanpur Wildlife Sanctuary	A1
IN-UP-06	Kudaiyya Marshland	A1, A4iii
IN-UP-07	Kurra Jheel	A1, A4iii
IN-UP-08	Lagga-Bagga Reserve Forest	A1
IN-UP-09	Lakh-Bahosi Bird Sanctuary	A1, A4iii
IN-UP-10	Narora	A1, A4iii
IN-UP-11	National Chambal Wildlife Sanctuary	A1, A4iii
IN-UP-12	Nawabganj Bird Sanctuary	A1, A4iii
IN-UP-13	Parvati Aranga Wildlife Sanctuary	A1, A4iii
IN-UP-14	Patna Bird Sanctuary	A1, A4i, A4iii
IN-UP-15	Payagpur (Baghetal) Jheel	A1, A4iii
IN-UP-16	Saman Bird Sanctuary	A1, A4i, A4iii
IN-UP-17	Samaspur Bird Sanctuary	A1, A4i, A4iii
IN-UP-18	Sandi Wildlife Sanctuary	A1, A4i, A4iii
IN-UP-19	Sarsai Nawar Lake	A1, A4i, A4iii
IN-UP-20	Sauj Lake	A1, A4i, A4iii
IN-UP-21	Sheikha Jheel	A1, A4i, A4iii
IN-UP-22	Sohagi Barwa Wildlife Sanctuary	A1
IN-UP-23	Soheldev Wildlife Sanctuary	A1
IN-UP-24	Sur Sarovar Bird Sanctuary	A1, A4iii
IN-UP-25	Surha Taal Wildlife Sanctuary	A1, A4i, A4iii

SANJAY KUMAR

The Egyptian Vulture (adult above) is in Endangered category of IUCN. It is widespread in the state, although in depleted numbers

Distribution of threatened birds in the state with IBA site code		
Common Name	Scientific Name	IBA Site Code
Critically Endangered		
White-backed Vulture	*Gyps bengalensis*	IN-UP-02, 04, 10, 14, 22, 23, IN-DL-01
Long-billed Vulture	*Gyps indicus*	IN-UP- 04, 10, 14, IN-DL-01
Slender-billed Vulture	*Gyps tenuirostris*	IN-UP-02, 04
Red-headed Vulture	*Sarcogyps calvus*	IN-UP-02, 04, 05, 08, 09, 11, 23
Bengal Florican	*Houbaropsis bengalensis*	IN-UP-02, 04, 05, 08
Endangered		
Baer's Pochard	*Aythya baeri*	IN-DL-01
Lesser Florican	*Sypheotides indica*	IN-UP-02
Black-bellied Tern	*Sterna acuticauda*	IN-UP-09, 10, 14, 17, IN-DL-01
Egyptian Vulture	*Neophron percnopterus*	IN-UP- 02, 03, 04, 05, 09, 10, 11, 12, 13, 14, 16, 17, 18, 19, 20, 21, 22, 23,24, 25, IN-DL-01
Vulnerable		
Swamp Francolin	*Francolinus gularis*	IN-UP-02, 03, 04, 05, 08, 22, 23
Marbled Teal	*Marmaronetta angustirostris*	IN-UP-02
Sarus Crane	*Grus antigone*	IN-UP-01, 02, 03, 04, 05, 06, 07, 08, 09, 10, 11, 12, 13, 14, 15, 16, 17, 18, 19, 20, 21, 22, 23,24, 25, IN-DL-01
Pallas's Fish-eagle	*Haliaeetus leucoryphus*	IN-UP-02, 04, 09 10, 11, 17, IN-DL-01
Greater Spotted Eagle	*Aquila clanga*	IN-UP-02, 03, 04, 07, 09, 10, 11, 12, 14, 16, 17, 21, 24, IN-DL-01
Indian Spotted Eagle	*Aquila hastata*	IN-UP-11, IN-DL-01
Lesser Adjutant	*Leptoptilos javanicus*	IN-UP-02, 04, 05, 22, 24, IN-DL-01
Indian Skimmer	*Rynchops albicollis*	IN-UP-10, 11, IN-DL-01
Great Slaty Woodpecker	*Mulleripicus pulverulentus pulverulentus*	IN-UP-02,04, 05, 22, 23
Bristled Grassbird	*Chaetornis striata*	IN-UP 02, 03, IN-DL-01
White-throated Bushchat	*Saxicola insignis*	Not yet reported from any IBA
Yellow Weaver	*Ploceus megarhynchus*	IN-UP-03 (?), 07 (?), 08 (?), IN-DL-01
Near Threatened		
Falcated Duck	*Anas falcata*	IN-UP-02
Ferruginous Duck	*Aythya nyroca*	IN-UP-02, 09, 10, 14, 18, IN-DL-01, IN-DL-01
Lesser Flamingo	*Phoenicopterus minor*	IN-UP-14, 24
Painted Stork	*Mycteria leucocephala*	IN-UP-02, 05, 06, 09, 10, 12, 14, 15, 17, 18, 19, 20, 21, 23, 24, IN-DL-01,
Black-necked Stork	*Ephippiorhynchus asiaticus*	IN-UP-02, 05, 06, 09, 10, 12, 14, 15, 16, 17, 18, 19, 20, 21, 22, 24, IN-DL-01

Distribution of threatened birds in the state with IBA site code (*contd.*)		
Common Name	Scientific Name	IBA Site Code
Near Threatened (*contd.*)		
Black-headed Ibis	*Threskiornis melanocephalus*	IN-UP-02, 04, 09, 10, 11,12, 14, 17, 18, 21, 24, IN-DL-01
Spot-billed Pelican	*Pelecanus philippensis*	IN-UP-02, 09, 11, 14, 24
Oriental Darter	*Anhinga melanogaster*	IN-UP-01,02, 04, 09, 10, 11,12, 14, 17,18, 21, 24, IN-DL-01
Laggar Falcon	*Falco jugger*	No record from IBAs
Lesser Fish-eagle	*Ichthyophaga humilis*	IN-UP-02, 22, IN-DL-01
Grey-headed Fish-eagle	*Ichthyophaga ichthyaetus*	IN-UP-02, 04, 05, 21, 22,23, 24
Cinereous Vulture	*Aegypius monachus*	IN-UP-02,04, 23
Pallid Harrier	*Circus macrourus*	IN-UP-03 (?), 11, 24
Eurasian Curlew	*Numenius arquata*	IN-UP-01,03, 10, 11, 24, IN-DL-01
River Tern	*Sterna aurantia*	IN-UP-01, 02,03, 04, 05, 06, 07, 09, 10, 11, 12, 13, 14, 15, 16, 17, 18, 20, 24, 25, IN-DL-01
River Lapwing	*Vanellus duvaucelli*	IN-UP-03,04,10, 11, IN-DL-01
European Roller	*Coracias garrulus*	IN-UP-03, 11, 24
Great Pied Hornbill	*Buceros bicornis*	IN-UP-02, 04, 23, 24
Rufous-rumped Grassbird	*Graminicola bengalensis*	IN-UP-02, 04

IN-DL-01: In the book *Important Bird Areas in India* (Islam & Rahmani 2004), Okhla Bird Sanctuary, Code IN-DL-01, was included in the Delhi State. As half of Okhla Bird Sanctuary falls in UP, we are listing here the Threatened and Near Threatened birds found in this IBA also.

SANJAY KUMAR

Sarus is the State Bird of Uttar Pradesh. Nearly 60% of India's Sarus Crane are found in the state

Species such as Long-billed Vulture *Gyps indicus* (Critically Endangered), Red-headed Vulture *Sarcogyps calvus* (Critically Endangered), Egyptian Vulture *Neophron percnopterus* (Endangered), Lesser Adjutant *Leptoptilos javanicus* (Vulnerable), and Great Slaty Woodpecker (Vulnerable) also occur in Uttar Pradesh (UP) but they occur in other states/countries in much larger numbers.

THREATS AND CONSERVATION ISSUES

Uttar Pradesh is one of the most densely populated states in India with less than 7% area under forest cover. As mentioned earlier, the Gangetic plains have practically no forest cover left. The marshes and wetlands of the Gangetic

RAJAT BHARGAVA

Overfishing is one of the major threats to waterbirds as it depletes their food supply and creates disturbance

drainage system show a long history of stability in geological sense, thus many marsh-dependent species are found such as Striated Marsh Warbler or Grassbird *Megalurus palustris*, Bristled Grassbird *Chaetornis striata*, Rufous-rumped Grassbird *Graminicola bengalensis,* Yellow-bellied Prinia *Prinia flaviventris*, Swamp Francolin *Francolinus gularis*, Bengal Florican *Houbaropsis bengalensis*, and various ducks. Unfortunately, one of the duck species, the Pink-headed Duck *Rhodonessa caryophyllacea*, has become extinct, not due to any geological upheaval but due to human-related activities.

Eighteen out of 25 IBAs of Uttar Pradesh qualify on the basis of A4iii criteria (site known to hold, on regular basis, more than 20,000 waterfowl each year). Most of these wetlands are under tremendous pressure from fishing, overgrazing, cultivation, drainage, and pollution. Sometimes ill-conceived government plans are major threats to these wetlands, as is the case with the extremely important

Controlled burning of forests and grasslands is the usual management practice, but its long-term impact has not being studied

Sarus habitat in the Mainpuri-Etawah region. Another issue of concern is the large-scale diversion of wetlands and other waterbodies for Water Chestnut (*singhara*) cultivation after getting the *patta* (official permission) for fishing. The use of pesticides to increase productivity has adversely impacted the habitat of Sarus Crane and waterfowl.

The forests and grasslands of Dudhwa, Kishanpur, Katerniaghat, Pilibhit, Suhelwa, and Sohagi Barwa remain strong and vital reservoirs of *terai* biodiversity, and are important social and economic assets (Kumar *et al.* 2002). These PAs/IBAs are well-protected. However, the effect of annual grass burning and spread of invasive species such as *Tiliacora acuminata* are not fully known. Changes in river hydrology, associated siltation, and excessive ground water exploitation are causing changes in forest and grassland composition and structure (Kumar *et al.* 2002). Encroachment of forest land is still a major issue and vital corridors are still being lost (e.g., Dudhwa and Katerniaghat, Garha corridor between Mala and Deoria range of Pilibhit R.F. and Dudhwa and Kishanpur). Livestock grazing is a major problem, especially in sanctuaries.

General Recommendations

Besides the recommendations that we have given for each threatened species, we have general recommendations for bird protection in Uttar Pradesh. These generic recommendations are not comprehensive but indicative. We agree that much more work has to be done to secure the long-term survival of birdlife of the state.

(1) Studies on the impact of pesticides on birds should be started immediately with the help of Sálim Ali Centre for Ornithology and Natural History (SACON) as they have been doing such studies. Local universities should be involved.

ASAD R. RAHMANI

Grass harvest under the strict supervision of the Forest Department can provide much-needed resources and also create suitable habitat for species such as Bengal Florican

Stealing of fire wood, ostensibly for personal use, is a major socioeconomic issue in many PAs of the state

(2) Annual bird census in important PAs and IBAs should be taken up, involving experts.

(3) Ban on veterinary use of diclofenac should be implemented fully with the help of vets and paravets. Regular monitoring of carcasses for diclofenac should be taken up by BNHS, WII, and other institutions in collaboration with the forest department.

(4) Non-protected IBAs should be protected under the Wildlife Protection Act as sanctuaries or conservation/community reserves.

(5) Long-term and short-term research on threatened bird species should be promoted by the Government of Uttar Pradesh.

(6) Latest technology such as satellite tracking, DNA sampling, use of unmanned aerial vehicles (UAV) or Conservation Drone should be used for research and management.

(7) Large-scale bird ringing and colour banding should be started in the state with the help of professionals from BNHS and other organizations.

(8) Capacity building workshops for the frontline staff of the forest department should be held regularly. They should be funded by the UP government.

(9) Regular monitoring of species such as Sarus Crane, Bengal Florican, Swamp Francolin, Pallas's Fish-eagle, vultures, Black-necked Stork, and other birds should be started with appropriate funding mechanism.

(10) Monitoring of common birds such as House Sparrow, Indian Roller, Hoopoe, Koel, shrikes, orioles, etc., should be started under Citizen Science Programmes, involving a large number of people.

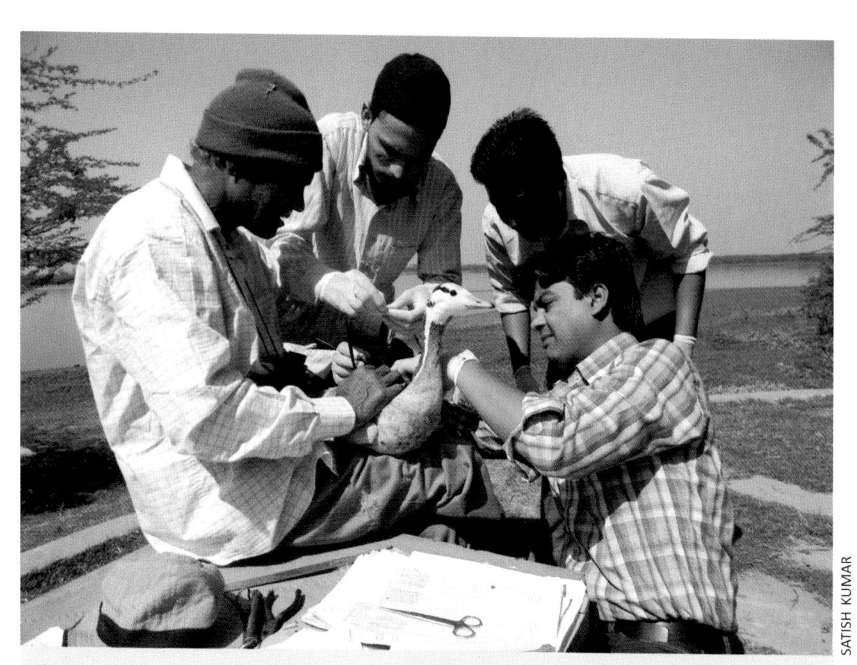
SATISH KUMAR

The UP Forest Department has a very enlightened policy of promoting wildlife research in the state. Above, researchers of BNHS and Department of Wildlife Sciences, Aligarh Muslim University putting PTT on a Bar-headed Goose in Sur Sarovar Bird Sanctuary

(11) Monitoring of waterfowl in important sanctuaries/wetlands such as Sandi, Nawabganj, Patna Jheel, Sur Sarovar, Lakh-Bahosi, Sheikha Jheel, Samaspur, Saman, Hakimpur, Sitadwar, and others should be started by the Forest Department with the help of NGOs and experts.

(12) Control of illegal bird trade and poaching throughout UP, especially for controlling organized trade in Swamp Francolin, Sarus Crane, Black-necked Stork, and waterbirds at various underground bird markets.

(13) Indiscriminate use of pesticides in the cultivation of Water Chestnut in natural wetlands should be controlled by appropriate measures.

(14) Fishing contracts in most of the wetlands create all sorts of disturbance and destruction to these habitats and their biodiversity.

(15) The directive of the Supreme Court regarding protection of ponds/waterbodies from encroachments is a valuable judgement. If the state government implements it in letter and spirit, it will be a remarkable achievement. A round-the-clock Waterbodies Protection Cell may be created at state level, for receiving complaints directly and effective monitoring of the disposal of complaints.

(16) It may be suggested to the state government that before erection of electric poles and high voltage wires near IBAs/PAs or other important non-protected IBAs, a No Objection Certificate may be obtained from the local

PAs and IBAs such as Nawabganj Bird Sanctuary play a very important role in conservation education of the general public

DFO, so that no developmental activity adversely impacts the habitat or the birds.

(17) A district-level wetland atlas based on remote sensing data and a perspective plan for conservation of the wetlands should be made for every district. The district wetland committee should be made active and made the nodal body to take up conservation of these waterbodies. The Sitapur model of wetland conservation using existing schemes and the community is an excellent example in the state, which can easily be emulated statewide (Kumar & Srivastav 2011). Under this conservation model, some basic activities like erection of permanent boundary pillars, desilting (where required) and de-weeding may be taken up. Local village wetland management committees may be constituted to look after the wetlands.

(18) An award could be instituted for individuals/organizations doing outstanding work on various threatened species to encourage such work.

(19) A biennial compendium of threatened species could be published and made available in the public domain, which includes data about census if any, habitat, conservation issues, efforts of the state government, and future plan of action.

(20) An interpretation centre with interactive models, audio-visual display of threatened birds may be proposed at Lucknow Zoo or Dudhwa NP for awareness generation among the masses.

References

Ali, S. and Ripley, S.D. (1987) *Compact Edition of the Handbook of India and Pakistan.* 2nd edn. Oxford University Press, Delhi.

BirdLife International (2001) *Threatened Birds of Asia: BirdLife International Red Data Book*. BirdLife International, Cambridge, UK.

BirdLife International (undated) *Important Bird Areas (IBAs) in Asia: Project briefing book.* BirdLife International, Cambridge, UK. Unpublished.

BirdLife International (2013) Species factsheets. Downloaded from http://www.birdlife.org.

Islam, M.Z. and Rahmani, A.R. (2004) *Important Bird Areas in India: Priority Sites for Conservation*. Indian Bird Conservation Network: Bombay Natural History Society and BirdLife International, UK. Pp xviii + 1133.

Javed, S. and Rahmani, A.R. (1998) Conservation of the avifauna of Dudwa National Park, India. *Forktail* 14: 55–64.

Kumar, H., Mathur, P.K., Lehmkuhl, J.F., Khati, D.S., De, R. and Longwah, W. (2002) *Management of Forests in India for Biological Diversity and Forest Productivity: A New Perpective: Terai Conservation Area (TCA).* Vol. VI. WII-USDA Forest Service Collaborative Project Report. Wildlife Institute of India, Dehra Dun. Pp. 158.

Kumar, S. and Srivastav, N. (2011) *Conservation of Potential wetlands in district Sitapur*. U.P. State Biodiversity Board. Pp. 143.

Ministry of Environment and Forests (2001) *Status of Forest of India*. Forest Survey of India, Dehra Dun.

Rahmani, A.R. (1989) Status of the Black-necked Stork *Ephippiorhynchus asiaticus* in the Indian subcontinent. *Forktail* 5: 99–110.

Rahmani, A.R. (2012) *Threatened Birds of India – Their Conservation Requirements*. Indian Bird Conservation Network: Bombay Natural History Society, Royal Society for the Protection of Birds and BirdLife International. Oxford University Press. Pp. xvi + 864.

Rahmani, A.R., Narayan, G. and Rosalind, L. (1990) Status of the Greater Adjutant (*Leptoptilos dubius*) in the Indian Subcontinent. *Colonial Waterbirds* 13(2): 139–142.

Rodgers, W. and Panwar, H.S. (1988) *Planning a Protected Area Network in India*. Vols 1 & 2. Wildlife Institute of India, Dehradun.

Sundar, K.S.G. and Kaur, J. (2001) Distribution and nesting sites of the Black-necked Stork *Ephippiorhynchus asiaticus*. *JBNHS* 98(2): 276–278.

Urquhart, E. (2002) *Stonechats: A guide to the Genus* Saxicola. Christopher Helm, London. Pp. 320.

■ ■ ■

Updates in the IUCN Red List of Threatened Birds 2013

BirdLife International is the official agency responsible for preparing the IUCN Red List of globally threatened birds. Earlier this was done every three to four years, but now the List is revised every year. Final changes in the IUCN Red List of birds for 2013 were declared in November 2013 and subsequently released on the BirdLife International website. The revised Red List of threatened birds of India now has 16 Critically Endangered, 18 Endangered, 53 Vulnerable, 75 Near Threatened, and 03 Data Deficient species. Thus a total of 165 species recorded in India are listed in the IUCN Red List of 2013.

The following important updates have been made in the Red List for India: the species in bold are found in Uttar Pradesh.

The following two species have been uplisted from Vulnerable to Endangered:

i. Manipur Bush Quail *Perdicula manipurensis* Not found in UP
ii. **Yellow-breasted Bunting *Emberiza aureola*** Recently reported in UP

The following five species have been uplisted from Least Concern to Near Threatened:

i. **Great Stone-Plover (Thick-knee) *Esacus recurvirostris***
ii. **Alexandrine Parakeet *Psittacula eupatria***
iii. Grey-headed Parakeet *Psittacula finschii* Not found in UP
iv. Rosy-headed Parakeet *Psittacula roseata** Not found in UP
v. **Red-breasted Parakeet *Psittacula alexandri*** Not found in UP (?)

The Blossom-headed Parakeet* (*Handbook* Nos. 557 to 559, Ali & Ripley 1987) was divided into two species with two subspecies each: Northern Blossom-headed Parakeet *Psittacula cyanocephala bengalensis*, Southern Blossom-headed Parakeet *Psittacula cyanocephala cyanocephala*; and Assam Blossom-headed Parakeet *Psittacula roseata roseata*, Arakan Blossom-headed Parakeet *Psittacula roseata juncae*. Rasmussen & Anderton (2005, 2012) named the two species as Plum-headed Parakeet *Psittacula cyanocephala* and Rosy-headed Parakeet *Psittacula roseata*. Grimmett *et al*. (2011) have retained the original name Blossom-headed Parakeet for *Psittacula roseata*. Rosy-headed or Blossom-headed Parakeet is not found in UP, and Plum-headed Parakeet *Psittacula cyanocephala* is a Least Concern species, so both are not described in this book.

The reasons why they have been uplisted/added are given below:

1. **Yellow-breasted Bunting *Emberiza aureola*** Yellow-breasted Bunting has been uplisted from Vulnerable to Endangered because of indications that the rate of population decline is more rapid than previously thought. It is now listed as Endangered, despite its high local abundance, because of compelling evidence that it is undergoing a very rapid population decline owing mainly to trapping in its non-breeding range. However, scientific monitoring is required to assess the decline in its wintering habitats (BirdLife International 2014).

Yellow-breasted Bunting breeds across the northern Palaearctic from Finland, Belarus, and Ukraine in the west, through Kazakhstan, China, and Mongolia, to far-eastern Russia, Korea, and northern Japan. In the autumn, birds stop over in large numbers to moult in the Yangtze Valley, China, before proceeding to their winter quarters.

RAJAT BHARGAVA

This bunting winters in a relatively small region in South and Southeast Asia, which includes eastern Nepal, north-eastern India, Bangladesh, Myanmar, southern China, Cambodia, Laos, Vietnam, and Thailand (BirdLife International 2014). In India, it is mainly found in the Northeast and northern West Bengal, with one record from Ambala. It is found in Nepal, but not specifically reported from Uttar Pradesh. Javed & Rahmani (1998) have mentioned this species as occasional in Dudhwa National Park. However, recently in March 2014, more than 150 were seen in western Uttar Pradesh (Bhargava *et al.* in preparation).

2. **Great Stone-Plover (Thick-knee) *Esacus recurvirostris*** Previously known as *Burhinus recurvirostris*, this species has been uplisted from Least Concern to Near Threatened as it is expected to undergo a moderately rapid population decline over the next three generations, owing to human pressures on riverine ecosystems and the construction of dams. It has already undergone steep declines in Southeast Asia, but its status currently appears more secure in India.

Great Stone-Plover occurs across a wide range in southern Asia, being found in Iran, Pakistan, India, Sri Lanka, Nepal, Bhutan, Bangladesh, Myanmar, Thailand, Laos, Cambodia, Vietnam, and southern China. According to Pravin J., it remains more numerous on the Indian subcontinent, where it prefers larger rivers though it also occurs on still water, but declines are believed to have taken place here too (BirdLife International 2014). In Uttar Pradesh, it is found on all large rivers but its population remains unknown.

SANJAY KUMAR

3. Alexandrine Parakeet *Psittacula eupatria* This species has been uplisted from Least Concern to Near Threatened on the basis of new information about its population trend. It is listed as Near Threatened because it is suspected to be undergoing a moderately rapid population decline owing to ongoing habitat loss, persecution, and trapping (BirdLife International 2014).

RAJAT BHARGAVA

Alexandrine Parakeet is widespread in South and Southeast Asia, ranging from Pakistan, through most of India (including the Andaman Islands and Narcondam Island), Sri Lanka, Nepal, Bhutan, Bangladesh, Myanmar, Thailand, Laos, Cambodia, and Vietnam (BirdLife International 2014). In Uttar Pradesh, it is fairly common but the population trend is unknown.

4. Red-breasted Parakeet *Psittacula alexandri* Red-breasted Parakeet has been uplisted from Least Concern to Near Threatened on the basis of new information about its population trend. It is suspected to be undergoing a moderately rapid population decline owing to ongoing trapping pressure, persecution, and habitat loss.

RAJAT BHARGAVA

Red-breasted Parakeet occurs in South and Southeast Asia, from northern and eastern India (including the Andaman Islands), Nepal, Bhutan, Bangladesh, Myanmar, Thailand, Laos, Cambodia, Vietnam, and southern China, with populations in Indonesia and some nearby islands. It is said to be easily seen in the foothills and adjoining plains of the Himalaya from Uttarakhand to Arunachal Pradesh, India. It was not reported by Javed & Rahmani (1998) from Dudhwa.

Trade of parakeets in India

In India, the biggest threat to parakeets is from bird trade. Thousands of parakeets belonging to nearly all species found in India are captured to cater to the demands of the pet bird market. Since hundreds of years, people in India have been keeping parakeets as pets mainly because they are easy to keep, live long, and have no special needs. This has resulted in an organized illegal trade in parakeets across the country. Pet trade in all Indian species is totally banned. This threat remains for the following three species of parakeets which are not so widely distributed: Nicobar Parakeet *Psittacula caniceps*, Long-tailed Parakeet *P. longicauda*, and Derbyan Parakeet *P. derbiana* were already listed as Near Threatened, with illegal trade posing the most significant threat. With the recent update in the Red List, four more species of parakeets have joined the Near Threatened category. These are Alexandrine *P. eupatria*, Grey-headed *P. finschii*, Blossom-headed or Plum-headed *P. roseata*, and Red-breasted Parakeet *P. alexandri*.

Hundreds of parakeets are collected and traded annually in the country. They are taken from the wild and smuggled to various parts of India and beyond. The bulk of the trade is in three to four week-old chicks. Parakeets are caught using nets and bird-lime. Adult parakeets are traded throughout the year, with chicks arriving in the trade between December and June. For every bird that reaches the market place, several are believed to die en route (http://www.wwfindia.org website, accessed on March 16, 2014).

Of the 12 native species of parakeets found in India, eight are regularly found being traded illegally. They include the Alexandrine, Rose-ringed, Plum-headed, Red-breasted, Malabar, Himalayan, and Finsch's (Grey-headed) Parakeet and Vernal Hanging-parrot (http://www.wwfindia.org website, accessed on March 16, 2014).

ABRAR AHMED

Parakeets are some of the most popular traded species in India despite an official ban since the early 1990s

According to Ahmed (2011), ornithologist and bird trade consultant to TRAFFIC India, the Alexandrine Parakeet is one of the most sought after species in the Indian live bird trade and is traded in large volumes throughout the year. The chicks are collected from forested areas of Himachal Pradesh, Jammu, and Punjab, and transported to bird markets in Delhi, Mumbai, Hyderabad, Patna, Lucknow, and Kolkata. Many specimens are smuggled by Indian dealers via Pakistan, Nepal, and Bangladesh to bird markets in various parts of the world. TRAFFIC India, with support from WWF-India, has been doing exemplary work in creating awareness about the illegal trade of parakeets in India.

According to Abrar Ahmed (*pers. comm.* 2013), who has done extensive field work on bird trade in India, the estimated number of parakeets in the Indian bird trade are: Rose-ringed Parakeet: 100,000–150,000; Plum-headed Parakeet 30,000–40,000; Slaty-headed or Himalayan Parakeet: 200–300; Red-breasted Parakeet: 8,000–10,000; Alexandrine Parakeets: 25,000–30,000 birds.

This estimate was made in 2000–2001. With such numbers of parakeets being supplied to the bird trade, no wonder the species are declining in their natural habitats.

References

Ahmed, A. (2011) Trouble in parrot paradise. *TRAFFIC Post* 12: 11–12.

Ali, S. and Ripley, S.D. (1987) *Compact Edition of the Handbook of the Birds of India and Pakistan*. Oxford University Press, New Delhi.

Bhargava, R. (2012) *Live Bird Trade in India*. Ph.D. Thesis submitted to the University of Mumbai, Mumbai. Pp. 508.

Bhargava, R., Rahmani, A.R. and De, R. (in preparation) Recent sighting of Yellow-breasted Bunting *Emberiza aureola* in Suhelva Wildlife Sanctuary, Uttar Pradesh.

BirdLife International (2014) IUCN Red List for birds. Downloaded from http://www.birdlife.org on 16/03/2014.

Grimmett, R., Inskipp, C. and Inskipp, T. (2011) *Birds of the Indian Subcontinent*. 2nd edn. Oxford University Press, London.

http://www.wwfindia.org website, accessed on March 16, 2014.

Javed, S. and Rahmani, A.R. (1998) *Conservation of the avifauna of Dudwa National Park, India. Forktail* 14: 55–64.

Rasmussen, P.C. and Anderton, J.C. (2005) *Birds of South Asia: The Ripley Guide*. Vols 1&2. Smithsonian Institution Press, Washington D.C. and Lynx Edicions, Barcelona, Spain.

Rasmussen, P.C. and Anderton, J.C. (2012) *Birds of South Asia: The Ripley Guide*. Revised edition. Vols 1&2. Smithsonian Institution Press, Washington D.C. and Lynx Edicions, Barcelona, Spain.

■ ■ ■

Extinct Birds of Uttar Pradesh

The following three bird species have become extinct in Uttar Pradesh. Their extinction warns us that unless we take effective measures quickly, more species will be added to this unfortunate list. No state or country can be proud of such a list.

Pink-headed Duck *Rhodonessa caryophyllacea* (Latham 1790)

THE GAME BIRDS OF INDIA, E.C. STUART BAKER, Vol. 1

The Pink-headed Duck is a large bird (*c.* 60 cm) with a long neck and unique colour pattern. The male is dark brown above and below, with the bill, head, and neck pink, and a blackish throat. The head is partially tufted. Female duller and browner, has a pale greyish-pink head and upper neck, with a brown crown and hind neck and dull pink bill. It is easily confused with the male Red-crested Pochard *Netta rufina*, as the latter species also has a conspicuous red head (although the colour is actually very different from the Pink-headed Duck).

The Pink-headed Duck was formerly distributed in the Gangetic plains from central Uttar Pradesh, east to extreme west Assam, and south to east Orissa. There are records from Bangladesh and Myanmar, but rarely from Nepal. A combination of hunting and habitat loss exterminated this species. The last authenticated field sighting was in 1935 in Darbhanga, Bihar by C.M. Inglis.

Historical records of the Pink-headed Duck from Uttar Pradesh are from **Balia**, **Shahjahanpur**, **Lakhimpur-Kheri**, **Gonda**, **Rahimabad**, **Lucknow**, **Basti**, **Gorakhpur**, **Mohanlalganj**, and **Fatehpur** (BirdLife International 2001). There is a specimen of Pink-headed Duck in Lucknow Museum which is located in the Zoological Garden. Although the prospect of finding it in India is rather bleak as its original habitat is almost gone, nevertheless, some perennial backwaters of Girijapuri reservoir in Bahraich appear to be suitable, with shallow vegetated pools surrounded by dense forests.

Siberian Crane *Grus leucogeranus* Pallas 1773

The Siberian Crane is perhaps the most famous migratory bird in India. Unfortunately, it does not come to India anymore as the central population which used to migrate to India is considered to be extinct. The last recorded observation of the Siberian Crane in Keoladeo National Park, Rajasthan was in 2002. Not many people know that it was fairly common in the wetlands of western Uttar Pradesh, particularly Mainpuri-Etawah areas, where A.O. Hume collected many specimens in the 19th century. According to BirdLife International (2001), the Siberian Crane was seen or collected at the following sites in Uttar Pradesh: One collected from a flock of 25 birds, in the winter of 1858–1859, many seen at different jheels in **Etawah** in the winter of 1865–1866 and 1866–1867; **Suman** jheel, **Etawah**, there are three specimens in British Museum (Natural History) collected in February 1871; **Indugarh** jheel, seen in January 1871; five seen in a shallow jheel near **Sandila**, undated; three seen on four occasions, February 1859 in **Sandi** jheel; some seen on four occasions, December 1859 in **Hilgee**, Chowka river, Avadh; some specimens collected from **Fatehgarh** jheel, in February 1873, December 1974/February 1875 and December 1876; and **Payagpur** jheel in Bahraich, seen around 1937. Payagpur is the last reported sighting of the Siberian Crane from Uttar Pradesh. As this migratory bird is extinct in India, it is unlikely that it will appear again in Uttar Pradesh.

OTTO PFISTER

Great Indian Bustard *Ardeotis nigriceps* (Vigors 1831)

The Great Indian Bustard is now a Critically Endangered bird with a population estimated at 200 birds, mainly in Rajasthan, Gujarat, Maharashtra, Karnataka, and Andhra Pradesh. Nearly a hundred years ago, it was widespread at least in the western half of Uttar Pradesh. Even up to the 1980s, there were chances of sighting it in Jhansi, Lalitpur, Banda, Etawah, and Agra districts, as some birds were surviving in the adjoining Shivpuri and Gwalior districts of Madhya Pradesh. However, since their near disappearance from Madhya Pradesh by the year 2000

MOHAN LAL MEENA

(Rahmani 2012), the chance of Great Indian Bustard occurring in the geographical limits of Uttar Pradesh is now nil.

Earlier, it was seen or shot in Roorkee (now in Uttarakhand), **Deoband**, **Muzaffarnagar**, **Garhmukteshwar**, **Lakhimpur-Kheri**, **Agra**, **Lucknow**, **Orai**, and **Jhansi** (Rahmani & Manakadan 1990, BirdLife International 2001). A nest was located in **Mirzapur** district in August 1943 (Lowther 1949). Even now, some parts of Mirzapur district are suitable for this species, but with such a low global population it is very unlikely that the Great Indian Bustard will ever be found again in the state.

References

BirdLife International (2001) *Threatened Birds of Asia. BirdLife International Red Data Book*. 2 vols. BirdLife International, Cambridge, UK.

Lowther, E.H.N. (1949) *A Bird Photographer in India*. Oxford University Press, London.

Rahmani, A.R. and Manakadan, R. (1990) The past and present distribution of the Great Indian Bustard *Ardeotis nigriceps* (Vigors). *JBNHS* 87: 175–194.

Rahmani, A.R. (2012) *Threatened Birds of India – Their Conservation Requirements*. Indian Bird Conservation Network, Bombay Natural History Society, BirdLife International, and Royal Society for the Protection of Birds. Oxford University Press, Mumbai. Pp. xvi + 864.

■ ■ ■

Threatened birds likely to be found in Uttar Pradesh

The following three species are likely to be found in Uttar Pradesh as two were reported in the past, and one could be a passage migrant. Two being small and uncommon, could be easily missed. Moreover, there is a general lack of good birdwatchers in the state.

Greater Adjutant
Leptoptilos dubius (Gmelin 1789)

At one time a very common species in eastern and north-east India, the Greater Adjutant is now an Endangered species due to massive decline in its numbers. Probably less than 1,000 adult birds now remain in India, mainly found in Bihar and Assam. Earlier it was reported from Pakistan through northern India, Nepal, and Bangladesh, to Myanmar, Thailand, Laos, Vietnam, and Cambodia. Till the early 1980s, some stray birds were seen in Keoladeo National Park in Rajasthan and Bhindawas Bird Sanctuary in Haryana (Rahmani 2012), but now we do not have any confirmed record from these sites.

In Uttar Pradesh, it was reported from **Agra**, **Lucknow**, **Loni**, and **Gorakhpur** (BirdLife International 2001). A nest, presumably of this species, was located at Mansur Ghat, north Gorakhpur district, in December 1861 (Beavan 1865–1868). As this bird wanders around after breeding is over, it is likely that it may appear again in the territorial boundary of Uttar Pradesh. Its cousin, the Lesser Adjutant *Leptoptilos javanicus* is regularly seen in the state.

White-browed Bushchat *Saxicola macrorhynchus* (Stoliczka 1872)

The White-browed Bushchat, earlier known as Stoliczka's Whinchat, an arid and semi-arid habitat specialist, is poorly known, but is thought to have a small, declining population as a result of agricultural intensification and encroachment, which qualifies it as Vulnerable (BirdLife International 2013). The immature bird and female White-browed Bushchat look similar to the female of Common Stonechat *Saxicola torquatus*, except for the more conspicuous white eyebrow and longish bill. The adult breeding male is however quite distinct — dark above with mostly blackish mask and wings, broad white supercilium and band along inner wing-coverts and mostly white primary coverts, whitish underparts, usually with a buffy-yellow tinge across the breast. Non-breeding male has broad buffish fringing above and buffish-fringed remiges, with less white. Female resembles non-breeding male but lacks the dark mask and white on tail, and has duller wings. Juvenile is darker brown above than the female, with buff streaks and spots and whitish below, and indistinctly brown-mottled throat and breast.

DHRITIMAN MUKHERJEE

Historically, the White-browed Bushchat was reported from Punjab, Haryana, Uttar Pradesh, Rajasthan, and Gujarat, adjacent parts of Punjab and Sind in Pakistan, and even up to Afghanistan (BirdLife International 2001). It is still found in Rajasthan and Gujarat, with some confirmed reports from Maharashtra and Haryana (Rahmani 2012). Ali & Ripley (1987) have mentioned a record from **Aligarh**. However, Rahmani (1993, 1996) could not find any individual in the district despite many searches, even though its semi-arid habitat is still present in some parts of the district. With records from Hissar and Gurgaon in Haryana, and an earlier record from Aligarh, it is likely to be present in the territorial limits of Uttar Pradesh. More intensive and extensive searches are required.

Tytler's Leaf-warbler *Phylloscopus tytleri* Brooks 1872

Tytler's Leaf-warbler is a small (10 cm), drab olive and dull-brown bird, difficult to distinguish in the field from similar-sized and similar-looking *Phylloscopus* species. In fresh plumage, it is olive above and whitish below, but when it is worn, it becomes greyish above and dingy below. It breeds in a very limited area of western Himalaya from northeastern Afghanistan (Nuristan) (Paludan 1959) eastwards at least through Kashmir (Ali & Ripley 1987), and winters in the Western

Ghats from Maharashtra to Kerala. During migration, it possibly passes through Nepal, so it may also pass through Uttar Pradesh. We need good recorders to know its true status in the state. This species has a moderately small population, which is suspected to be declining as a result of habitat loss and degradation on both the breeding and wintering grounds. Therefore, it is classified as Near Threatened (BirdLife International 2013).

VINAYAK YARDI

References

Ali, S. and Ripley, S.D. (1987) *Compact Edition of the Handbook of the Birds of India and Pakistan*. Oxford University Press, New Delhi.

Beavan, R.C. (1865–1868) Notes on various Indian birds. *Ibis* (2)1: 400–423; (2)3: 430–455; (2)4: 73–85, 165–181, 355–356, 370–406.

BirdLife International (2001) *Threatened Birds of Asia. BirdLife International Red Data Book*. 2 vols. BirdLife International, Cambridge, UK.

BirdLife International (2013) Species factsheets. Downloaded from http://www.birdlife.org.

Paludan, K. (1959) On the birds of Afghanistan. *Vidensk. Medd. Dansk Naturhist. Foren.* 122: 1–332.

Rahmani, A.R. (1993) Little known Oriental Bird: White-browed Bushchat. *OBC Bulletin* 17: 28–30.

Rahmani, A.R. (1996) Status and distribution of Stoliczka's Bushchat *Saxicola macrorhyncha* in India. *Forktail* 12: 61–77. Published 1997.

Rahmani, A.R. (2012) *Threatened Birds of India – Their Conservation Requirements*. Indian Bird Conservation Network, Bombay Natural History Society, BirdLife International and Royal Society for the Protection of Birds. Oxford University Press, Mumbai. Pp. xvi + 864.

■ ■ ■

HISTORY OF AVICULTURAL TRADITIONS AND THE UTILISATION AND TRADE OF WILD BIRDS IN UTTAR PRADESH

Rajat Bhargava

Bird keeping and trading in India has an ancient history. Parakeets and mynas have been popular in royal courts, caged and kept as a source of entertainment. Since time immemorial, pigeons have been used as couriers and have even shaped the course of many a battle. Other winged creatures such as junglefowl have been reared by humans to cater to the dinner table. Evidence of this close relationship between man and birds is found in several medieval Indian paintings. Popular Indian folklore has it that the parakeet and myna were the model romantic pair. Not surprisingly, many poets composed lyrics on this entrancing duo, and artists captured the pair on canvas. These two birds which have an uncanny ability to mimic human voices were also a great favourite of Mughal emperors.

The Great Mughal Akbar was said to be very fond of birds and kept them for their song, exotic coloration, and companionship. He also kept gamebirds like partridges and quails, which were used for bird fights. And, of course, no royal court in those days was complete without the falcon, used as it was to hunt gamebirds. Akbar's biography *Ain-i-Akbari* gives graphic details of all the birds in his court. The manuscript is not only of great use to historians, but also provides ornithologists a look at the bird world of the time, giving detailed accounts of at least 13 raptor species. Take this sample from Akbar's chronicles by historian Abul Fazl Allami: "From eagerness to purchase and from inexperience, people pay high sum for falcons. His Majesty [Akbar] allows dealers very reasonable profits; but for motive of equity he has limited the prices. The dealers are to get their gain, but buyers ought not to be cheated." At that time, the Goshawk was considered the most superior bird, and was sold for 12 *muhr*s (8–10 gm gold coins), while common species like the Shikra cost a quarter rupee. Dealings were done on the basis of bird moults at the time of sale. The *Ain-i-Akbari* also refers to the Mirshikars, a clan dealing in birds, who were used by Akbar for falconry, as well as for trapping and hunting of animals. Even today, this clan is found in eastern India, they eke out a living by trapping and selling birds and animals. *Ain-i-Akbari* describes how this clan used indigenous techniques to trap birds. Indian religious texts too have several references to birds. For instance, in both the *Ramayana* and the *Mahabharata*, there are stories about pet parakeets. The two epics also make reference to the Hindu Baheliya (bird trapper) tribe. And in *Ramayana*, several birds play a central part in the story – *Jatayu* (a vulture) for instance.

Black Francolin singing competition

Perhaps the parakeet is the most revered pet bird in the country. Indeed, the Rose-ringed Parakeet *Psittacula krameri* and the Alexandrine Parakeet *P. eupatria* can be found in Hindu temples where they are taught to chant the Hindu god Rama's name. Parakeets are kept as companions or as mascots for good luck. Incidentally, Alexander the Great is supposed to have taken them to Greece. Besides being pets, parakeets are believed to have divine powers. Even today, in the age of internet and computer-generated horoscopes, the streetside astrologer with his caged parrots can be seen in many towns and cities (Ahmed 1997).

It is interesting to see how history has shaped the bird trade. When the English arrived in India, they made it fashionable to own exotic species such as budgerigars *Melopsittacus undulatus*, cockatiels *Nymphicus hollandicus*, and lovebirds *Agapornis* spp. as pets. Till 1972, when the Indian WildLife (Protection) Act came into force, there was unregulated trade in wild flora and fauna in India, and India was the biggest exporter of live birds to all continents, with Japan being the largest importer followed by European countries (Inskipp 1975, 1983). A study conducted by the Royal Society for Protection of Birds between 1971 and 1974 at Heathrow Airport, London revealed that every year India alone exported several million birds of nearly 275 species (Inskipp 1975).

In 1990–1991, the Government of India banned export and domestic trade in all native wild birds in India. More than two decades after the ban, the practice and culture of wild-caught bird keeping has become more or less 'near-extinct', or the trade in wild birds for food has gone underground. As there is no restriction on the sale and rearing of exotic (foreign) birds within the country, most bird markets in India now sell domesticated exotic species. However, except for studies on bird trade by TRAFFIC India/WWF-India documenting more than 450 species of native birds trapped for several uses (Ahmed 1997, 1999, 2004, 2010, 2012), there is little documentation of avicultural practices across India.

Traditionally, every area in the country has its own history of bird keeping or utilisation that varies from place to place in terms of species preference, or a certain tribe that traded in a particular group of birds. The degree of bird exploitation and use is in direct correlation to the available forest type and habitat in a particular region or state. The greater the diversity of birds and availability of forest, the higher the density of tribes dealing with birds and bird related cultural diversity.

Among the states of India, Uttar Pradesh has a significant cultural and bird trade history, being second only to Kolkata, West Bengal. Uttar Pradesh was a major hub of traditional bird trapping tribes, markets, and export centres. Some of the well-known markets in Uttar Pradesh are the Nakhas Chowk market in Lucknow, Baheliya Toli in Varanasi, Phasiyana Mohalla in Jhansi, Nakhas Kona in Allahabad, Mirshikar Toli in Gorakhpur, along with smaller markets in Moradabad, Meerut, Pilibhit, Bareilly, Unnao, and Kanpur. Among the former bird export centres of India, the largest were in Meerut and Rampur, thanks to their close proximity to Delhi Airport. Birds were brought to these centres from all over the country, especially the softbills from the Himalayan foothills. Other birds were brought

RAJAT BHARGAVA

Bulbuls were trained to fight each other

from Assam by Bihar and West Bengal bird dealers/trappers. Endemic birds such as the Malabar Parakeet *Psittacula columboides* were brought from far flung places such as Alvaee in Kerala. Species with major populations in south India, such as the Black-headed Munia *Lonchura malacca*, were brought from Chennai and Bengaluru, but the trapping was done by visiting trappers from Lucknow. Waterbirds were locally caught during winter from the rich wetlands. Agra is the world's biggest collection centre for peacock tail feathers (Ahmed 2008).

Uttar Pradesh is home to a number of bird trapping tribes who have now been deprived of their traditional vocation, especially the Baheliyas, a knowledgeable community of Hindu bird trappers. Clusters of this clan scattered

RAJAT BHARGAVA

Racket-tailed Drongo was a much admired pet in olden times

throughout Uttar Pradesh are so significant that according to a note prepared as a supplementary answer to a Lok Sabha question regarding trade of wild birds, the Ministry of Environment & Forests (8th June, 1998) based on information from the State Chief Wildlife Wardens revealed that there were 191 licensed dealers and trappers prior to the ban on trade in Uttar Pradesh. Furthermore, there were 32,000 families involved in trapping and trading of birds in Uttar Pradesh alone, the highest number compared to any other Indian state, with

RAJAT BHARGAVA

Streetside astrologer with caged parakeets

400 families reported from West Bengal. As the present state of Uttarakhand was formerly part of Uttar Pradesh, these numbers also include families from Uttarakhand. The art of trapping softbills such as Shama *Copsychus malabaricus*, Red-billed Leothrix *Leiothrix lutea*, Orange-billed Leafbird *Chloropsis hardwickii*, and similar species by mimicking hill birds by this clan has vanished since the past decade (Bhargava 1995).

Other bird trapping tribes in Uttar Pradesh are the Pathamis or Jabjalies, Muslim tribals who were originally fishermen. In some places, especially Lucknow, they are skilled cage-makers. In eastern UP, a few Mirshikar families live in places such as Gorakhpur, Mirzapur, Varanasi, Gonda, and Bahraich, that dominate bird trade. They are skilled trappers of waterbirds, especially waders and waterfowl.

Keeping in mind the number of people involved in bird trade, one can imagine the magnitude of bird trade and the number of birds and species captured in UP during the pre-ban era. With stricter enforcement and increasing awareness about the cruelty and numbers of wild birds exploited through trading and capture, bird trade is currently restricted to domesticated exotic cage birds, nevertheless, underground trade in native wild birds for food, merit release by religious people, for black magic and sorcery still exists in the state, but in comparatively negligible numbers.

As more and more exotic birds are in demand with open sales across UP and neighbouring states, the foreign bird keeping culture is a boon for bird breeders across the nation and a number of full-time breeders are involved in breeding

Captive Sarus

colour mutations of foreign birds. The bulk of the breeding stock comes from Kolkata and Hyderabad, and locally bred birds are easily available too (Ahmed 1997).

It is important to document the diminishing traditional bird practices in UP which are almost extinct not only in this state, but practically all over India. Most of these bird practices were popularised during the reign of the Nawabs of Lucknow. Bird sport events were mastered by a handful of people who had the knowledge, and this was passed on from one generation to the next by the practicing *ustads* (experts). Events such as *Bazdari*, *Kabutar-bazi*, *Teetar-bazi*, *Bater-bazi* and *Bulbul-bazi* were a livelihood and passion for a certain class of people whose lives revolved around these sports. *Bazdari* or falconry was a very popular sport in India. Falcons and hawks were trained to hunt smaller birds and sometimes mammals. Some very interesting events in the state of Uttar Pradesh have been mentioned by the late falconer S.M. Osman in his book *Hunters of the Air – a falconer's notes*. For instance, Osman (1991) writes "Pitting a *Bubo bubo* against a Peregrine Falcon *Falco peregrinus* was considered 'top-notch sport' as Americans would put it, during the heyday of Indian falconry. This was during the 1870s, when all princely families, *thakurs*, *zamindars*, and Colonel Radcliff, and both cavalry and infantry officers maintained establishments of trained falcons. In fact, these events were rather prestigious, and the Governer General would, on occasion, grace the gala affair. The event usually took place in *khaddar*, not far from Meerut." Falconry today is virtually an extinct sport in UP.

In the *terai* belt of western UP, sub-adult males of Black Francolin were acclimatised and trained for singing competitions. Birds that kept calling till the end were judged the winners. Bushlarks and Shama participated in singing competitions where these birds were judged on their songs and musical notes.

In Lucknow, Allahabad, Kanpur, and surrounding areas, partridge and quail fights were popular before the ban on bird keeping (Ahmed 1997). Mainly captive-reared Grey Francolin *Francolinus pondicerianus* were trained and made to fight in public prior to their breeding season. Similarly, Rain Quail *Coturnix chinensis* were trained to fight against each other. Chukor *Alectoris chukar*, brought from Nepal and the Himalayan foothills, were trained to fight each other, apart from being popular pets.

ABRAR AHMED

In small towns on the Indo-Nepal border, derivatives from threatened species such as vultures and hornbills (see highlighted section) are advertised by traditional medicinal practitioners

Before the onset of spring, Redvented Bulbul *Pycnonotus cafer* males were trained to fight each other. People of all castes and religions gathered on such occasions. For two to three months, the participants were into nothing except caring for their prizefighters. Once the competition was over, the bulbuls were released into the wild.

The keeping of Red Avadavat *Amandava amandava* as pets due to their beautiful coloration and song has been popular throughout Uttar Pradesh, and Red Avadavat fights were quite rare (Bhargava 2012). Today there is hardly an Indian alive who can train and organize such traditional bird events. The Baya Weaver *Ploceus philipinus* was trained to perform small tricks, while parakeets and sometimes munias were trained to pick cards for street performances. Until recently, the Kalandar tribes around Agra were masters in training the large horned-owls (*Bubo* spp.) for street performances to "purify" amulets (Ahmed 2010).

Because of the cost, place, and time involved in such practices, combined with the conservation movement in India, and the total ban on bird trade, the bird trapper is a much despised individual, scorned by conservationists, animal rights activists, and law enforcement officials. However, bird trappers and keepers have vast traditional knowledge, which can be put to alternative uses in the service of nature conservation. The Bombay Natural History Society set a splendid example when Mirshikar tribals from Bihar, including the legendary Ali Husain, were inducted into the BNHS Bird Migration Project, to which they have contributed with their traditional bird trapping skills. Thus the provision of respectable and economically viable livelihood has been ensured for several former trappers.

The hobby of birdwatching and photography, and wildlife research is gaining prominence in India. However, there is lack of research on the stakeholders that were dependent for their livelihood on birds and related activities. Anthropological studies on traditional tribes attached with birds and most importantly alternative rehabilitation schemes for traditional bird trapping tribes need to be examined. Cultivating a "birds and people" culture, along with the documentation of avicultural practices, including the art and craftsmanship of cage-making and trapping techniques, is also the need of the hour. Uttar Pradesh, a culturally rich place for such bird practices, can even host a museum on such fading bird practices.

References

Allami, Abul Fazl (1981) *The Ain-i-Akbari*. M.M. Publisher Pvt. Ltd., New Delhi. (Translated from the original Persian by H. Blochmann)

Ahmed, A. (1997) *Live Bird Trade in Northern India*. WWF/TRAFFIC-India, New Delhi. Pp. 104.

Ahmed, A. (1999) *Fraudulence in the Indian live bird trade: An identification monograph for control of illegal trade*. TRAFFIC-India/WWF-India, New Delhi. Pp. 25.

Ahmed, A. (2004) Illegal Bird Trade. In: Islam, M.Z. and Rahmani, A.R. *Important Bird Areas in India: Priority sites for conservation*. Indian Bird Conservation Network: Bombay Natural History Society and BirdLife International. Pp. 66–70.

Ahmed, A. (2008) *The trade in Peacock tail feathers in India* (A rapid survey). (TRAFFIC-INDIA, WWF-INDIA) New Delhi. Pp. 28.

Ahmed, A. (2010) *Imperilled Custodians of the Night: A study on the illegal trade, trapping and use of owls in India*. TRAFFIC India/WWF-India, New Delhi. Pp. 76.

Ahmed, A (2012) Trade in Threatened Birds in India. Pp. 40–72. In: Rahmani, A.R. *Threatened Birds of India – Their Conservation Requirements*. Indian Bird Conservation Network: Bombay Natural History Society, Royal Society for the Protection of Birds and BirdLife International, UK. Oxford University Press, Mumbai.

Bhargava, R. (1995) Trapping Hill Birds – A Vanishing Art. *Newsletter for Birdwatchers* 35(6): 102–104.

Bhargava, R. (2012) *Birds of Meerut*. 509 ASC Battalion, Meerut Cantt. Pp. 95.

Inskipp, T. (1975) *All Heaven in a Rage. A Study of Importation of Wild Birds into the United Kingdom*. Royal Society for the Protection of Birds. Pp. 41.

Inskipp, T. (1983) The Indian Bird Trade. *TRAFFIC Bulletin* V (3/4): 26–46.

Osman, S.M. (1991) *Hunters of the Air – a Falconer's Notes*. WWF-India, New Delhi.

■ ■ ■

White-backed or White-rumped Vulture

Gyps bengalensis (Gmelin 1788)

ASAD R. RAHMANI

Uttar Pradesh is or was one of the most important states of India for the White-backed Vulture, where up to the late 1980s millions of birds were found. Subsequently, the introduction of the killer drug diclofenac for veterinary use led to a catastrophic decline from the early 1990s, and the species declined so rapidly that BirdLife International (2001) listed is as Critically Endangered. Despite conservation efforts, its status has not changed and perhaps it is much more threatened now than it was in the 1990s. According to BirdLife International (2013), it qualifies as Critically Endangered because it has suffered an extremely rapid population decline, primarily as a result of feeding on carcasses of animals treated with the non-steroidal anti-inflammatory drug (NSAID) diclofenac.

Field Characters: The White-backed Vulture is the smallest (85 cm) of all Gyps vultures, and appears to be the basal from which other species of this clade of genus *Gyps* have diverged (Seibold & Helbig 1995, Johnson *et al.* 2006). It weighs 9 lb to 13 lb, measures 89–93 cm in length, and has a wingspan of 210–216 cm. It is mainly dark blackish brown with naked, thick, dark brown neck, with a white ruff at the base of the neck, dark silvery upper mandible, and conspicuous white rump (visible in flight or when spreading its wings). In flight, a broad whitish

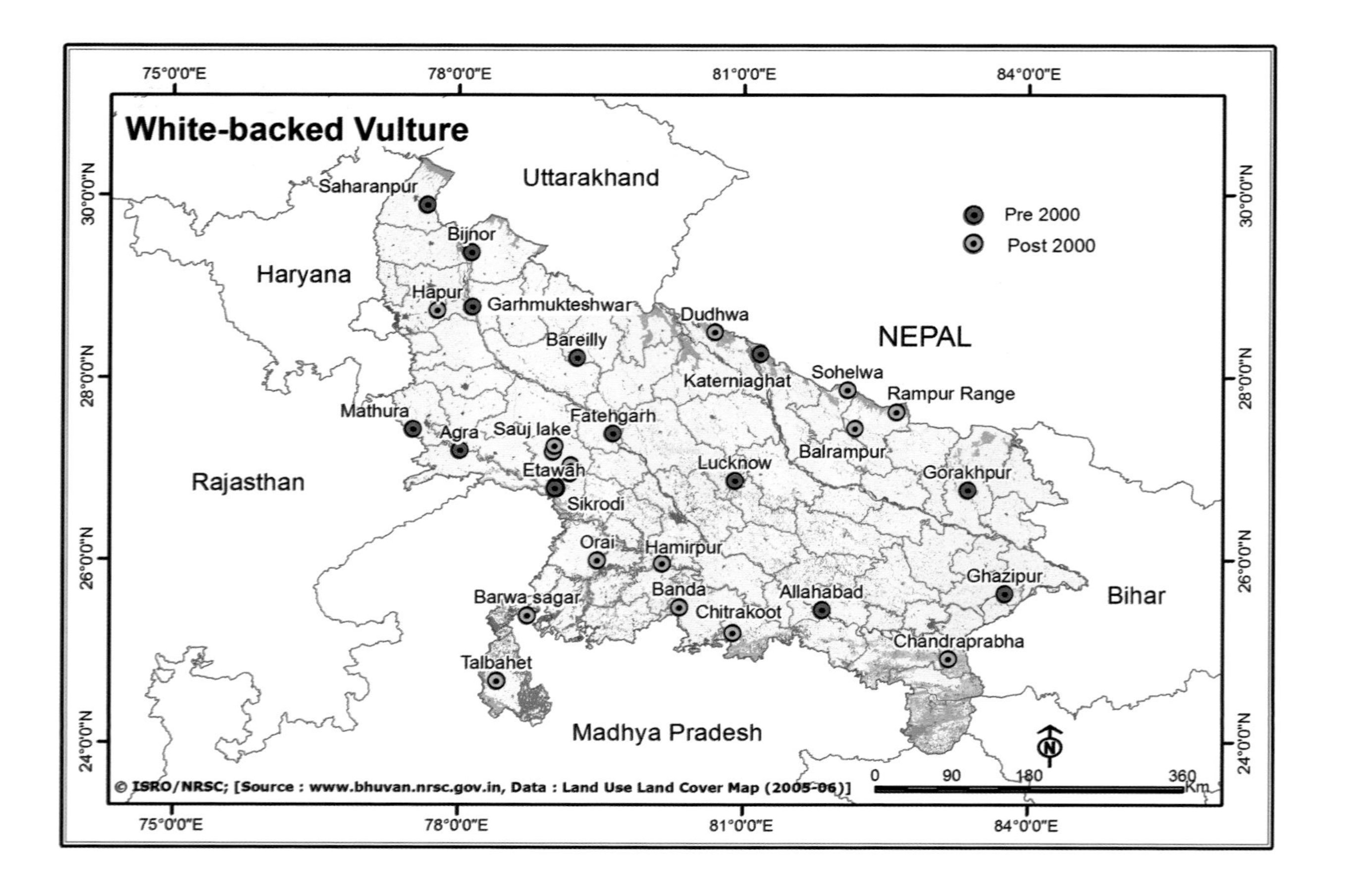

White-backed Vulture
Pre 2000
Post 2000
Uttarakhand
Haryana
Rajasthan
NEPAL
Bihar
Madhya Pradesh
Saharanpur
Bijnor
Hapur
Garhmukteshwar
Dudhwa
Bareilly
Katerniaghat
Sohelwa
Rampur Range
Mathura
Agra
Sauj lake
Fatehgarh
Balrampur
Etawah
Lucknow
Gorakhpur
Sikrodi
Orai
Hamirpur
Ghazipur
Barwa sagar
Banda
Allahabad
Chitrakoot
Chandraprabha
Talbahet
75°0'0"E
78°0'0"E
81°0'0"E
84°0'0"E
30°0'0"N
28°0'0"N
26°0'0"N
24°0'0"N
0 90 180 360 Km
© ISRO/NRSC; [Source : www.bhuvan.nrsc.gov.in, Data : Land Use Land Cover Map (2005-06)]

band along the underside of the wings is characteristic. Immatures are more brownish than black, and without white rump or underwing bands. Head and neck covered with dirty white fluffy down. According to Ali & Ripley (1987), "Impossible to distinguish with certainty in the field from Long-billed Vulture, with which it is commonly associated over most of its range."

Distribution: Before the 1990s, the White-backed was probably the most abundant vulture in the world, particularly in the northern states of India. It was also reported from Pakistan, Bangladesh, Nepal, Bhutan, Myanmar, Thailand, Laos, Cambodia, and South Vietnam, and earlier from southern China and Malaysia, but nowhere as abundant as in India, southern parts of Nepal and the Punjab province of Pakistan. It has been recorded from southeast Afghanistan and Iran where its status is currently unknown. According to BirdLife International (2001), it disappeared from most of Southeast Asia in the early 20th century, and the only viable populations in the region are now found in Cambodia (Pain *et al.* 2004) and Myanmar (both populations probably in the low hundreds).

Records from Uttar Pradesh: In Uttar Pradesh, the White-backed Vulture is now limited to certain pockets and found in limited, countable numbers. It is still found in small numbers in almost the whole of the Terai, and adjoining Nepal hills. In most of the recent sightings, we found mixed flocks of vultures that included Himalayan Griffon *Gyps himalayensis* and Eurasian Griffon *Gyps fulvus*, Slender-billed *Gyps tenuirostris* and a few Egyptian Vultures *Neophron percnopterus*. Here we give some recent records.

In April 2009, more than 30 vultures were found resting on a sandbar in River Girwa, just opposite the Boat Club in Katerniaghat Wildlife Sanctuary. Again on January 2012, 18 vultures, mostly White-backed, were observed on a leafless tree in a grassland in Katerniaghat range. Nesting has been recorded in **Katerniaghat** WLS, mainly on Silk Cotton trees in the forest edges on both sides of the Girwa river.

In April 2011, Amit Mishra reported a flock of 20 White-backed Vultures (with Egyptian Vultures) in an agriculture field near **Bhira** town, Lakhimpur-Kheri, feeding on a carcass.

The White-backed Vulture is regularly sighted in **Sohelwa** WLS. Some records are as follows: Mixed flock of 50+ birds (with Himalayan and Eurasian Griffons) seen on the Balrampur-Tulsipur road on March 1, 2009, feeding on a carcass. Mixed flock of over 100 birds recorded near Bankatwa, feeding on a carcass (Gurmeet Singh *pers. comm.* 2013). In December 2009, over 30 recorded roosting on a Mango and a dry Silk Cotton tree in a grassland near Motipur reservoir. In January 2012, over 70 birds, some flying, some roosting on a leafless tree in a dry stream bed in Navashahr beat in Tulsipur Range. Regular sighting of 20 to 60 vultures was reported from August 2010 to February 2011 in Sohelwa East and West ranges, mostly around Hathiakunda and Bhainsahi streams (Khem Bahadur *pers. comm.* 2013).

Nikhil Shinde and Rajat Bhargava of BNHS have seen two each in Poorvi Suhelwa Range (December 27, 2013) and Rampur Range (December 22, 2013) of **Soheldev** Wildlife Sanctuary located in Tulsipur *tehsil* of Balrampur district and Bhinga *tehsil* of Sharavasti district along the Indo-Nepal border.

In the Bundelkhand area of Uttar Pradesh, some White-backed Vultures are found throughout the Vindhyan range. In **Lalitpur** district, we have round the year sightings (barring monsoon months) of 25 to 30 individuals on the cliffs and trees, alongside Betwa river, which forms a boundary between Uttar Pradesh and Madhya Pradesh. Even if not sighted, their presence can be judged from the enormous streaks of droppings on the rocks. Another spot in Lalitpur range is an ancient temple and a cave known as Muchkund cave. Nesting of White-backed is also recorded from Marawara range, which is the biggest range in terms of area, on both sides of Dhasan river, which marks the boundary between Lalitpur in UP and Sagar district in Madhya Pradesh. Large flocks of over 100 birds were reported in Gothra and Lakhanjar beat of Marawara range. There are confirmed sightings from Girar beat of Marawara range of White-backed, Long-billed, and Egyptian Vultures feeding on carcasses (Rakesh Kumar Prajapati *pers. comm.* 2013). Regular vulture sightings from Sati Anusuya temple and hillocks around it in district **Chitrakoot** have been reported in newspapers and by naturalists. Roosting of *c.* 40 vultures on the ramparts of the famous Kalinjar Fort were reported by Sanjeev Sharma. He also noted four nests in 2012 (Sanjeev Sharma *pers. comm.* 2013).

In western Uttar Pradesh, **Hapur** was at one time the biggest stronghold of this species where even now a few birds exist. For example, in December 2012, two vultures were recorded on the Hapur-Bulandshahr road being chased by dogs. In mid-2012, one White-backed was sighted at Sobhapur in **Meerut** with two Egyptian Vultures. On the outskirts of **Saharanpur**, two birds were recorded in 2012. In mid-2013, three White-backed were sighted on the **Rampur-Moradabad** road (Rajat Bhargava *pers. obs.* 2013).

Vulture decline: The most catastrophic and rapid decline of the White-backed Vulture (and other related Gyps species) has been seen in South Asia. This decline was first reported in newspapers in the mid-1990s (see Rahmani 2008) and later confirmed scientifically at Keoladeo NP (Prakash 1999) and all over India (Prakash *et al.* 2003, Prakash *et al.* 2007). Similar steep declines were noticed in Nepal (Baral *et al.* 2005) and Pakistan (Gilbert *et al.* 2006).

For many years, virus-related disease(s) was considered the likely cause of this catastrophic population crash, but in 2003, a non-steroidal anti-inflammatory drug (NSAID) diclofenac was identified as the culprit (Oaks *et al.* 2004a, b; Green *et al.* 2004, Shultz *et al.* 2004). This drug is used as a pain killer for domestic livestock. If an animal dies within 2–5 days of ingestion of diclofenac and vultures feed on its carcass, they suffer renal failure causing visceral gout (Oaks *et al.* 2004a). The three Gyps species of vultures have declined by 97%–99% during the last 20 years. As diclofenac is widely used even now despite being officially banned, the rate of decline is nearly 50% per year of the remaining populations (Green *et al.* 2004). Modelling shows that vulture declines at the observed rates can be caused by the contamination of less than 1% of livestock carcasses with levels of diclofenac lethal to vultures. The proportion of adult vultures which die with symptoms of diclofenac poisoning is consistent with that expected if diclofenac were the sole cause of the recent rapid population declines (Green *et al.* 2004, Pain *et al.* 2008).

Despite 97–99% decline in the numbers of White-backed Vulture (and other related species), it is still seen in many areas in India, although in small numbers. People have become more conservation-conscious and vulture sightings are reported in newspapers and bird e-groups. However, this does not mean that their population is recovering. BNHS maintains a database of all these sightings.

Ecology: The White-backed Vulture inhabits open countryside, avoiding thick forests and wooded hilly areas. As it feeds on large carcasses, it has to locate them visually, so it soars regularly on thermals, covering vast areas of hundreds of square kilometres in a single day. It finds food either by its own sightings or by following other descending vultures and scavengers.

The White-backed Vulture lives in flocks and breeds on tall trees in loose scattered colonies, however, young birds may nest solitarily. When it was abundant, breeding colonies were found even in Lucknow, Aligarh, Meerut, Sitapur, Shahjahanpur, Kanpur, Allahabad, and many other large cities. Nests were also seen on avenue trees with heavy traffic below, and inside bustling towns and villages.

Threats: Like the other two Critically Endangered Gyps species found in Uttar Pradesh, the White-backed Vulture is in real danger of becoming extinct in another 5–10 years if diclofenac is not effectively and completely banned from veterinary use. It has been found that even if less than 1% of the cattle carcasses have a lethal dose of diclofenac, Gyps vultures will continue to die at the observed rate (Green *et al.* 2004). In a carcass sampling study conducted from May 2004 to June 2005 across 12 states (Senacha *et al.* 2008), it was found that 10.1% of the cattle carcass samples contained diclofenac sufficient to cause widespread mortality of vultures. After the official ban on veterinary use of diclofenac in India in May 2006, two cattle carcass sampling studies showed that both the prevalence and concentration of diclofenac had fallen markedly 7–31 months after the implementation of the ban, with the true prevalence in the third survey estimated at 6.5% (Cuthbert *et al.* 2011). Modelling of the impact of this reduction in diclofenac on the expected rate of decline of the White-backed Vulture in India indicates that the decline rate has decreased to 40% of the rate before the ban, but is likely to be still rapid (about 18% per year). Hence, further efforts to remove diclofenac from vulture food are still needed if the future recovery or successful reintroduction of vultures is to be feasible.

There are some other causes such as inadvertent poisoning from baits placed to kill other species, electrocution by collision with high tension wires, injury

Large numbers of vultures feeding on carcasses is now a rare sight in Uttar Pradesh

from kite threads (particularly seen in Gujarat during the annual Kite Festival) and chicks falling from nests, but these factors play a minor role.

Organochlorine pesticide was found in egg and tissue samples, varying in concentrations from 0.002 mg/g of DDE in muscles of vultures from Mudumalai to 7.30 mg/g in liver samples of vultures from Delhi. Dieldrin varied from 0.003 to 0.015 mg/g. These pesticide levels have not, however, been implicated in the decline (Muralidharan *et al.* 2008).

Another hypothesis is that they are affected by avian malaria (Poharkar *et al.* 2009), but this cannot explain such large-scale mortality in such a short time across South Asia (Ishtiaq 2009).

Conservation measures underway: All three Gyps species have been included in Schedule I of the Indian Wildlife (Protection) Act, 1972, since 2000. They are listed in CITES Appendix II and CMS Appendix II. BirdLife International and IUCN have listed them as Critically Endangered. In 2004, IUCN World Congress passed a BNHS/RSPB/BirdLife-sponsored resolution urging all the range states to ensure effective protection of Gyps vultures. An International South Asian Vulture Recovery Plan has been developed and is being implemented in India, Nepal, and Pakistan. This Plan suggests establishing a minimum of three captive breeding centres, each capable of holding 25 pairs. In India, conservation breeding centres have been established in Pinjore in Haryana, Buxa in West Bengal, and Rani RF near Guwahati in Assam by the RSPB and BNHS in collaboration with the state forest departments. Captive breeding efforts are ongoing and met with success when two chicks hatched in early 2007 at a breeding centre in Pinjore, Haryana. Since then, the vultures are breeding in increasing numbers, and even double clutching and artificial incubation have been successful. For more details, please visit www.bnhs.org and www.rspb.org.

Long-billed Vulture

Gyps indicus (Scopoli 1786)

SATYENDRA KUMAR TIWARI

The Long-billed Vulture *Gyps indicus*, sometimes known as Indian Vulture, is classified as Critically Endangered by BirdLife International and IUCN because it has suffered an extremely rapid population decline as a result of feeding on carcasses of animals treated with the drug diclofenac.

Field Characters: A large, robust vulture (*c.* 92 cm), with a conspicuous white neck-ruff and a long black neck with pale down feathers which are missing in the Slender-billed Vulture *G. tenuirostris*.

Distribution: The Long-billed Vulture is a semi-endemic bird of India, with a small population surviving in the Sind province of Pakistan, near the India-Pakistan border. In India, is it found from the Gangetic plains almost up to Tamil Nadu. Along with the White-backed, the Long-billed Vulture was one of the most common vultures of India, till the 1990s when diclofenac was introduced for veterinary purposes.

Records from Uttar Pradesh: In Uttar Pradesh, it is difficult to draw its distribution boundary. What we know or presume is that wherever the Slender-billed is found, the Long-billed is absent. So basically it is found south of the River Ganga, particularly in the districts of Jhansi, Lalitpur, Etawah, Mirzapur, and Varanasi.

Ecology: Its ecology is not very different from that of the White-backed Vulture, as both used to be present in large numbers in the open countryside, sometimes in villages, near cultivated areas, and in lightly wooded areas. Earlier when it was common, it was found near cities and towns, particularly on carcass dumps and around slaughterhouses. It is a scavenger and feeds almost entirely on carrion, often with the White-backed Vulture. Unlike the White-backed that nests on trees, the Long-billed nests almost exclusively in small colonies on cliffs

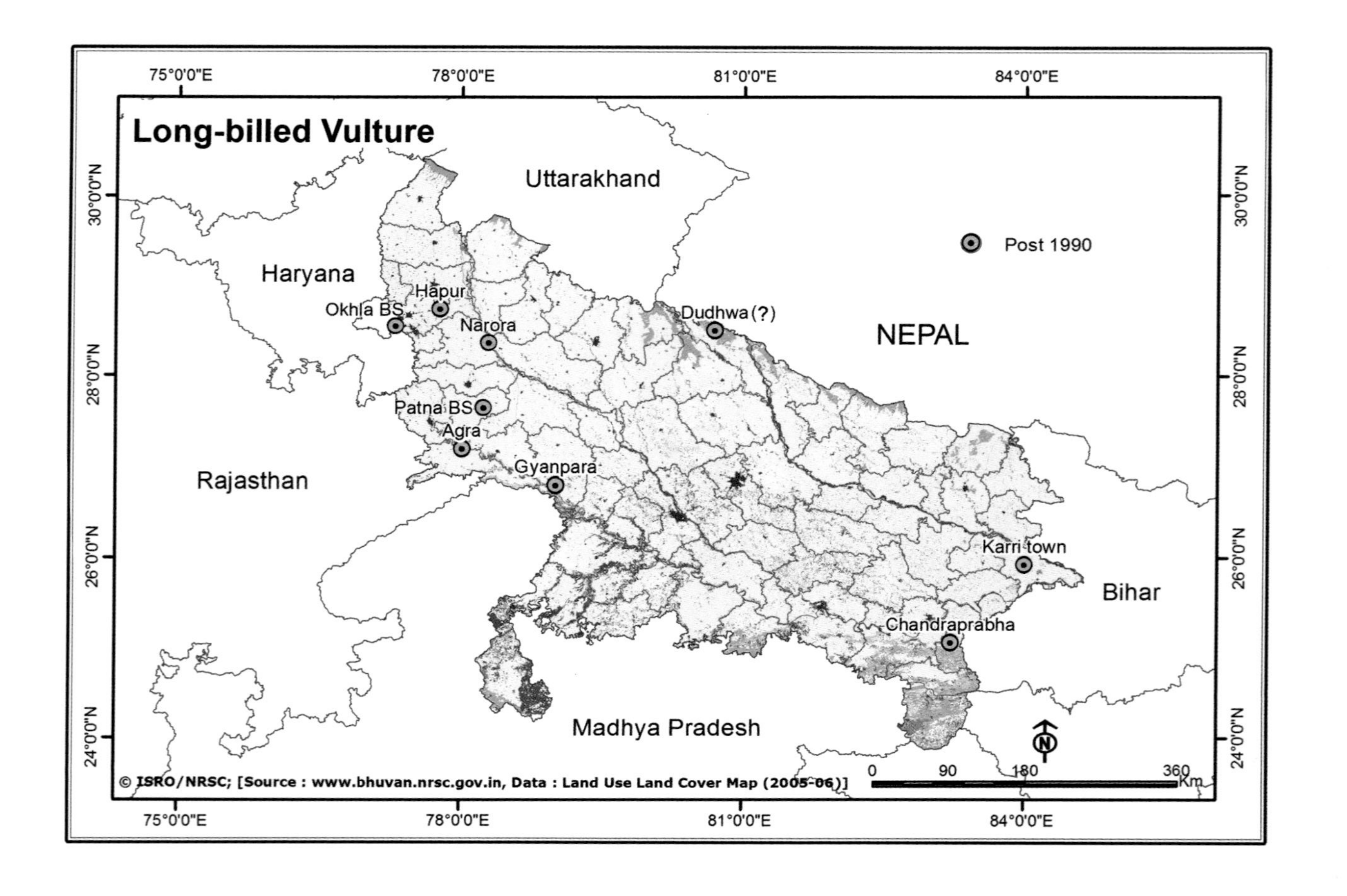
Long-billed Vulture
Uttarakhand
Haryana
Rajasthan
Madhya Pradesh
NEPAL
Bihar
Post 1990
Hapur
Okhla BS
Narora
Patna BS
Agra
Gyanpara
Dudhwa (?)
Karri town
Chandraprabha
75°0'0"E
78°0'0"E
81°0'0"E
84°0'0"E
30°0'0"N
28°0'0"N
26°0'0"N
24°0'0"N
0
90
180
360
Km
© ISRO/NRSC; [Source : www.bhuvan.nrsc.gov.in, Data : Land Use Land Cover Map (2005-06)]

and ruins. Where cliffs are absent, it sometimes nests on trees. During the day, it roams over hundreds of square kilometres in search of food, but returns to the same cliffs to roost at night. Nesting colonies are traditional and used year after year, clearly visible due to the white faecal markings below the nests.

Threats, Conservation Measures, and Recommendations: See Slender-billed Vulture.

Vulture Crisis

The extinction crisis that hit Asian vultures was first noticed by villagers in 1996–97 in western Uttar Pradesh. News started appearing in local newspapers of UP that vultures were disappearing. I remember reading in *Amar Ujala*, a newspaper published from Agra, that vultures had disappeared from the carcass dumps near Meerut. After reading the news, I kept wondering was happening to vultures, still very common in Aligarh and other areas. I started keeping a watch on vultures which still filled the skies over Aligarh. For some years, it was considered hearsay, or at best a local phenomenon, as there were still hundreds of thousands of vultures in our countryside. Dr. Vibhu Prakash, Deputy Director of BNHS who was working in Keoladeo National Park, Bharatpur at that time, confirmed that vultures were dying in large numbers. He had census data of the mid-1980s from Keoladeo, and when he compared his data with the figures for 1997–98, he got shocking results. Further confirmation came when he compared his all-India raptor survey data of 1990–93 with survey data of 2000–2003. The crisis was not local but pan-Indian. Till then, Gyps vultures, particularly the White-backed and Long-billed, were so abundant in north India that even if 50% were to disappear, we would still see thousands of them. Secondly, no one was monitoring them systematically in India to notice the declining trend.

Earlier it was thought that Gyps vultures were dying due to a viral disease, but in May 2003, scientists of The Peregrine Fund delivered a paper during the 'World Conference on Birds of Prey' in Budapest, where they clearly demonstrated that the painkiller diclofenac sodium was causing visceral gout in vultures that consumed livestock carcasses contaminated with the drug. Later, a paper was published on January 28, 2004 in *Nature*, conclusively proving that diclofenac sodium was the real culprit. I would say that this paper was the second turning point in our struggle to save the vultures. This paper with 13 authors belonging to six institutions, with Lindsay Oaks as the lead author, showed 100% correlation between the presence of diclofenac in vultures and kidney failure. After this paper, Vibhu and his team also analyzed vulture carcasses which we had meticulously preserved in deep freezes. We also found direct correlation between visceral gout and presence of diclofenac in vulture bodies. The evidence was all over the place – in research papers, in labs, in widespread dead birds, and in diclofenac ampoules in vet shops.

Diclofenac sodium as a painkiller was released for veterinary use in 1992–93 by the Government of India. Being cheap and very effective, it became extremely popular and soon its use was widespread.

The vulture crisis is a timely warning that we should start monitoring so-called common birds, as many of them are not so common anymore. Only large-scale scientific monitoring will provide early warnings to take timely conservation action. Many bird species have declined due to multiple factors such as habitat destruction and deterioration, pesticides, hunting, increase in stray dog populations, and increased urbanisation. Let us not wait for a crisis to overwhelm us before we take action.

Asad R. Rahmani

Slender-billed Vulture

Gyps tenuirostris Gray 1844

ASAD R. RAHMANI

The Slender-billed Vulture is perhaps the most threatened vulture in the world, with a very narrow distribution range, north of River Ganga and up to sub-Himalaya in North India, West Bengal, and east to Assam. Outside India, it was reported from Nepal, Bangladesh, Thailand, Malaysia, Laos, and Cambodia, and still occurs in Myanmar where a small population was recorded in Shan State (BirdLife International 2013). The Slender-billed Vulture's distribution does not overlap with the Long-billed Vulture, and unlike the latter, it nests on tall trees, sometimes near human habitation.

Field Characters: The Slender-billed Vulture is quite distinct from the Long-billed Vulture *Gyps indicus*, with its characteristic slender jet black neck, thin elongated bill, angular black head, long legs, and toes with dark claws at all ages. Head and neck appear almost naked, thickly creased and wrinkled (prominently visible at close quarters). The contour feathering on its lower body is loosely textured and sparse. Relatively small and sparse contour feathering on the leg gives rise to the conspicuous white down patch on the outer side of the leg. Overall, it has an unkempt, scruffy look. Identified in the field by its large and prominent ear canal, unlike that of the Long-billed Vulture and other Gyps species. It is similar in size to the Long-billed (92 cm), and distinctly larger than the White-backed (85 cm). Juveniles are very similar, but have a black head and neck with a hint of white down on the nape and upper neck.

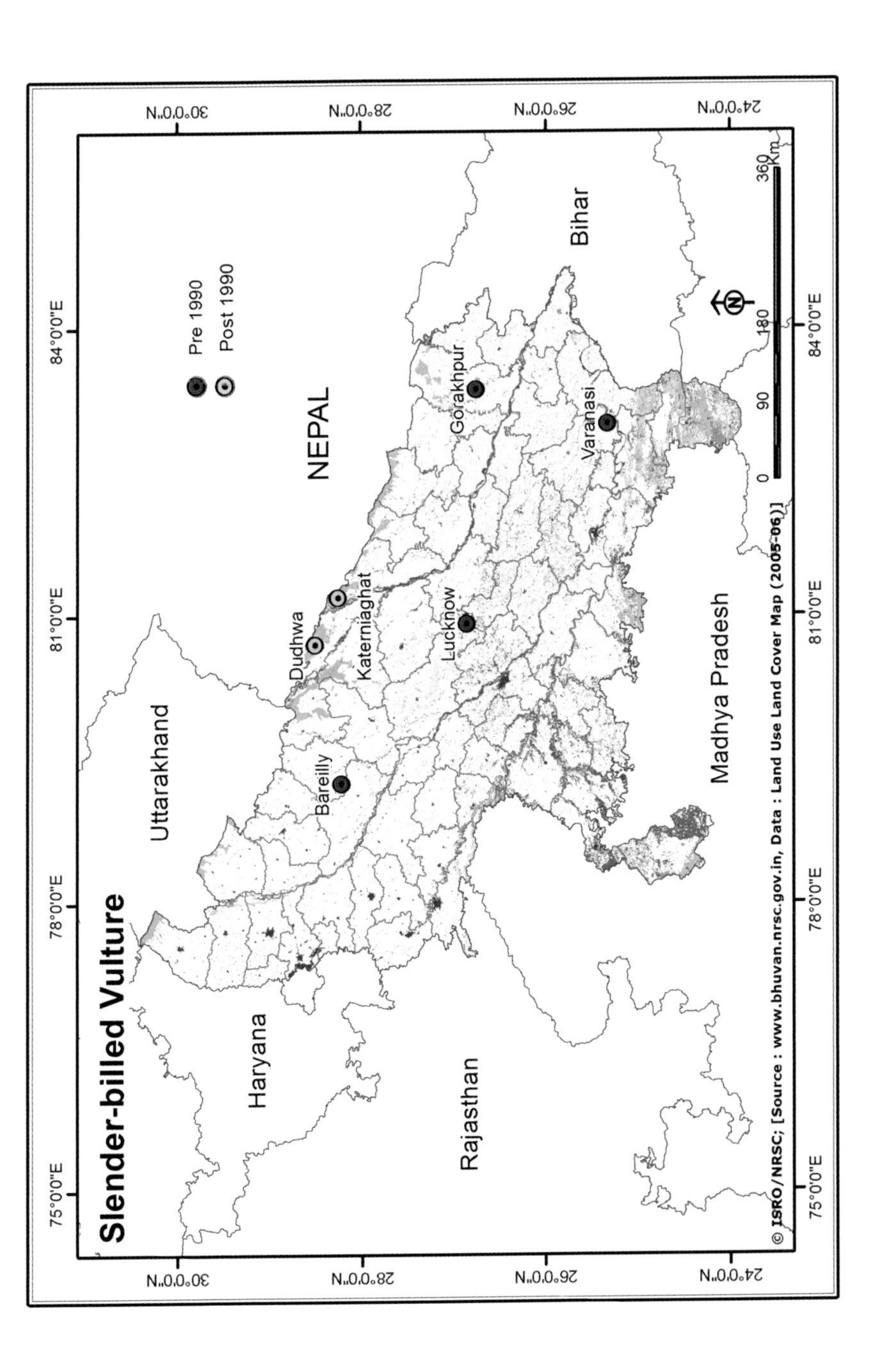

Slender-billed Vulture
Pre 1990
Post 1990
NEPAL
Uttarakhand
Haryana
Rajasthan
Madhya Pradesh
Bihar
Bareilly
Dudhwa
Katerniaghat
Lucknow
Gorakhpur
Varanasi
0
90
180
360
Km
75°0'0"E
78°0'0"E
81°0'0"E
84°0'0"E
30°0'0"N
28°0'0"N
26°0'0"N
24°0'0"N
© ISRO/NRSC; [Source : www.bhuvan.nrsc.gov.in, Data : Land Use Land Cover Map (2005-06)]

Distribution: Historically, the Slender-billed Vulture ranged throughout the Himalayan foothills of India, Nepal, north and central Bangladesh, Myanmar (except the north), and Southeast Asian countries, namely Thailand, Malaysia, Cambodia, and Laos.

It had already become very rare over most of its distributional ranges in Southeast Asia in the latter half of the 19th century and the first half of the 20th century, and now it has a very small, restricted distribution. In India and Nepal, the Slender-billed Vulture was common till the mid-1990s, but since the introduction of the non-steroidal anti-inflammatory drug (NSAID) diclofenac, it has suffered a massive decline along with the White-backed *Gyps bengalensis* and Long-billed Vultures (Prakash 1999, Green *et al*. 2004; Oaks *et al*. 2004a; Shultz *et al*. 2004; Swan *et al*. 2006a,b; Pain *et al*. 2008). The last estimate of population was below 1,000 individuals (Prakash *et al*. 2007).

Its present stronghold in India is mainly in the lower Himalaya and Gangetic plain from Himachal Pradesh and Haryana in the west, Uttar Pradesh, to southern West Bengal, and east through Assam and the Northeastern Hill states. Rahmani (2012) has given recent records in his book *Threatened Birds of India*, so we are not repeating them. Here we give records specific to Uttar Pradesh.

Records from Uttar Pradesh: In Uttar Pradesh and Uttarakhand, there were past records from Almora, Kumaon, **Bareilly**, Chakia, **Varanasi**, Dehradun, **Dudhwa** NP, **Gorakhpur**, Naini Tal, Pithoragarh, and Rajaji NP. However, presently it is observed only in **Dudhwa** and **Katerniaghat** and its adjoining area. It was photographed recently 11 km from Tanakpur near Banbasa in Uttarakhand, very close to the UP border, on April 2, 2010 along with White-backed Vulture and Eurasian Griffon *Gyps fulvus* on a horse carcass and on a tree. Nests have been located as recently as 2012 in Katerniaghat Sanctuary.

Ecology: Like the other *Gyps* species in India, the Slender-billed Vulture is an inhabitant of open dry country, often seen near human habitations, mainly at carcass dumps where it feeds, along with its congeneric species, on carrion. Despite its hooked bill and sharp claws, it does not kill its prey but feeds on carcasses of large or medium-sized ungulates. It tolerates human presence and sometimes breeds near villages on tall trees. It is a social bird — feeding, roosting, and resting in large loose flocks, often with other species of vultures.

The breeding season is in winter, from November to March. The nest is found on large tall trees, from 8–12 m, often in loose colonies. Only one egg is laid and both parents help in incubation and raising the chick.

Threats: The main threat to the Slender-billed and other species of Gyps vultures in Asia is from the veterinary use of diclofenac. For more details, see White-backed Vulture. Other contributory factors are changes in disposal method of dead livestock, unintentional poisoning and vehicle/train accidents, but these are probably of minor significance.

Conservation measures underway: Like the other two species of Gyps in India, the Slender-billed Vulture is included in Schedule I of the Indian Wildlife (Protection) Act, 1972, since 2000. It is listed in CITES Appendix II and CMS Appendix II. BirdLife International (2013) and IUCN have listed it as Critically Endangered since 2001. In 2004, the IUCN passed the BNHS/RSPB/BirdLife-sponsored resolution in its World Congress urging all the range states to ensure effective protection of Gyps vultures. A South Asian Vulture Recovery Plan (Anon. 2004) has been developed and is being implemented in India, Nepal, and Pakistan (this species is not found in Pakistan but the recovery plan is for all three Gyps species).

The Governments of India and Nepal have banned the use of diclofenac. *Ex situ* conservation efforts in India showed success in breeding this species for the first time in captivity, in 2008–2009.

RECOMMENDATIONS FOR THREE GYPS SPECIES OF VULTURES IN INDIA

The Government of India has officially banned the veterinary use of diclofenac and it is not available in the market. However, formulations of diclofenac for human use are widely used illegally on livestock despite the ban.

(1) The central government with the help of state governments should strictly implement the ban on veterinary use of diclofenac, including human-use formulations, on livestock.

(2) All competent organizations and agencies should seriously implement programmes to raise awareness of the hazards of diclofenac poisoning of vultures among the general public and especially among major stakeholders, including farmers, graziers, veterinarians, pharmacists, staff of government and state wildlife and agricultural agencies, religious and other groups which place special value on the continued existence of vultures.

(3) Appropriate authorities should undertake thorough evaluation of pharmaceuticals likely to be used in place of diclofenac to ensure that they are not toxic to vultures and other scavengers. The Ministry of Environment and Forests, Government of India should insist that any new NSAID for veterinary use should be safety-tested on vultures before it is introduced in the market.

(4) Other factors not responsible for the recent catastrophic declines may assume increasing significance in future, as the depleted populations fall still further. So all such factors including poisoning of cattle carcasses, injury or death of vultures due to kite flying (in Gujarat in particular), and disturbance to nests should be stopped by appropriate conservation measures.

(5) The Ministry of Environment and Forests should ensure that funding does not become a constraint in running the Vulture Conservation Breeding Programme, which is crucial for the survival of these valuable species that play a vital role in quick disposal of carcasses, and consequently in human environmental management.

A consortium Saving Asia's Vultures from Extinction (SAVE) was launched in February 2011 in Delhi and Kathmandu to provide a strategic framework through which the unprecedented problem and threat to South Asian Gyps vultures could be addressed across national boundaries. It provides a clear scientifically based outline of the priorities that need addressing to conserve the most threatened species, and also a recognized channel for supporters to ensure that resources are used to address those priorities.

SAVE consists of six core members: Bird Conservation Nepal, Bombay Natural History Society, International Centre for Birds of Prey (UK), National Trust for Nature Conservation (Nepal), the Royal Society for the Protection of Birds (UK), and WWF Pakistan, and a growing number of project and research partners including the Indian Veterinary Research Institute. Professor Ian Newton, world renowned raptor expert, agreed to take the chair for the first four years; and there are two main subcommittees that help drive the research, field actions, and advocacy that is needed.

Vulture conservation efforts in India are showing the first signs of success thanks to the initiative of the Indian government (led by the Ministry of Environment and Forests) in banning veterinary formulations of diclofenac in 2006 which has had an important impact in slowing the declines. The breeding programme includes three BNHS-run centres in Haryana, West Bengal, and Assam with the support of the respective state government Forest Departments, and the Central Zoo Authority is supporting further breeding facilities to extend these efforts at five more zoos. A Regional Steering Committee has been established through an IUCN initiative in 2012 with National Vulture Recovery Committees being set up in each of Pakistan, India, Nepal, and Bangladesh which will be an important forum for delivering the further measures required to conserve vultures in the Subcontinent.

A website, www.save-vultures.org provides full details and more information, as well as all key Asian vulture publications available for download, the manifesto, and most importantly a donations button where supporters can help ensure that resources are available to support these vital efforts.

www.save-vultures.org

Chris Bowden, SAVE Programme Manager

Red-headed or King Vulture

Sarcogyps calvus (Scopoli 1786)

DHRITIMAN MUKHERJEE

The Red-headed Vulture has suffered an extremely rapid population reduction in recent years, and it is predicted that this trend will continue, probably largely as a result of the birds feeding on carcasses of animals treated with the veterinary drug diclofenac, and perhaps in combination with other causes. For this reason it is classified as Critically Endangered (BirdLife International 2013).

Field Characters: The Red-headed is a medium-sized (76–86 cm) vulture, mainly black with bare red head, neck, and legs. Thigh patches and ruff are white. The male has yellowish eyes, while the female has dark reddish eyes. In flight, the conspicuous red head and legs, white breast, and white patches on the side of the thighs are diagnostic. Besides, the whitish band along the underwing lining is seen in flight. Immature birds are similar to adult, except they are overall brown; head pink and covered with white down; anterior flank and abdomen pale brown; posterior flank, abdomen, and undertail-coverts white, which is a diagnostic feature of the immature in flight.

Distribution: The Red-headed Vulture is found in South Asia and Southeast Asia, but always in low numbers. In India, it was widespread but never common (Ali & Ripley 1987, Grimmett *et al.* 1999, Rasmussen & Anderton 2005). Recent surveys (Cuthbert *et al.* 2006) indicate that in India it has undergone a rapid

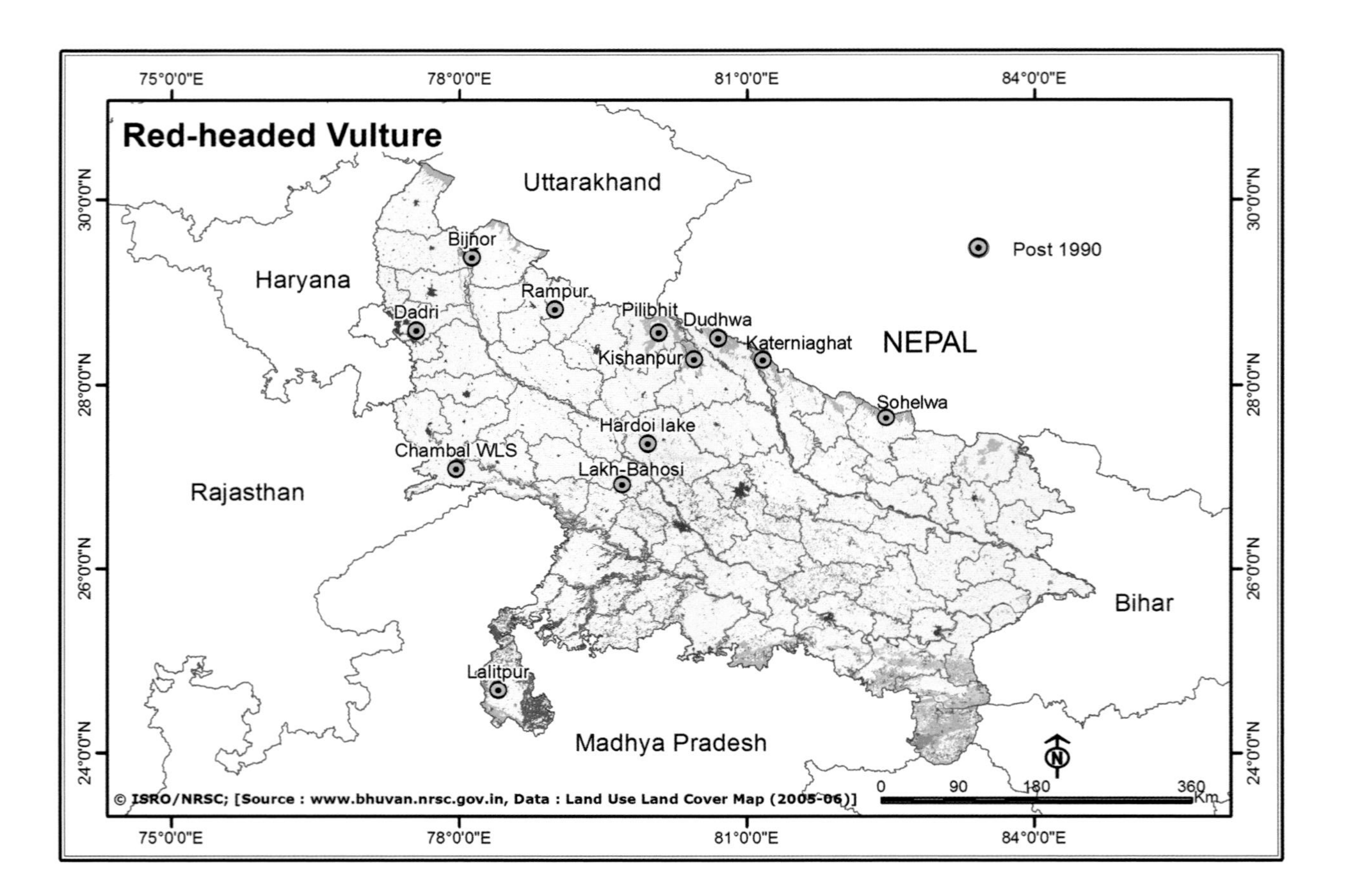

Red-headed Vulture
Uttarakhand
Haryana
Rajasthan
NEPAL
Post 1990
Bijnor
Dadri
Rampur
Pilibhit
Dudhwa
Kishanpur
Katerniaghat
Sohelwa
Hardoi lake
Chambal WLS
Lakh-Bahosi
Lalitpur
Madhya Pradesh
Bihar
75°0'0"E
78°0'0"E
81°0'0"E
84°0'0"E
30°0'0"N
28°0'0"N
26°0'0"N
24°0'0"N
N
0 90 180 360 Km
© ISRO/NRSC; [Source : www.bhuvan.nrsc.gov.in, Data : Land Use Land Cover Map (2005-06)]

population decline and is now rare or absent from some areas. Its recent distribution records from India are given by Rahmani (2012). Here, we describe records from Uttar Pradesh.

Records from Uttar Pradesh: It was reported in **Dudhwa** NP (Javed & Rahmani 1998) in the early 1990s, and even now it is occasionally seen (Sonu Rana *pers. comm.* 2013). Nikhil Devasar (*pers. comm.* 2013) has seen it in **Dadri** and **Dudhwa**. In August 2013, one individual was seen in flight on the outskirts of **Rampur** city towards Moradabad (Rajat Bhargava & Md. Akhlaq *pers. obs.*). Two individuals were sighted at **Bijnor** Barrage in February 2010. Earlier, it was occasionally seen around Hapur, but there are no recent sightings. It has also been reported from **Hastinapur** WLS in Meerut (Bhargava 2012). Kaajal Dasgupta (*pers. comm.* 2013) who has done extensive birding in Bareilly-Pilibhit-Shahjahanpur area, has seen three in flight near Sharda river in **Pilibhit** on May 28, 2013. Nikhil Shende of BNHS has seen one bird in flight on April 1, 2014 in Mahauf Range of **Pilibhit** forest.

Ecology: The Red-headed Vulture mainly inhabits dry deciduous forest and wooded hills, usually below 2,500 msl. It is found in open countryside and even in desert (outside Uttar Pradesh). A carcass would generally be attended by one or two Red-headed Vultures, mainly feeding on the soft tendons and small bones. Earlier, 20–30 individuals were seen, but now 6–8 individuals have been reported on a single carcass. It is mainly a carrion eater, but also hunts small prey occasionally. Its large bald head and neck enable it to reach deep into the cavities of a large carcass without soiling its feathers. The nesting season is from December to May; both sexes share parental duties. Incubation period is *c.* 80 days and complete fledging about four months from the date of hatching.

Threats: As it is a shy bird found in low numbers, its decline has not caught the attention of conservationists or the government, unlike the Gyps vultures. Though there is currently no direct evidence to link the decline in this species with diclofenac poisoning, the geographic extent and rate of decline are very similar to the decline in the Gyps populations for which the impact of diclofenac poisoning is now established. Counts of Red-headed Vulture carried out in 13 protected areas in India from 1991–1993 were repeated in 2000 and revealed a significant decline of around 48% (Prakash *et al.* 2003). Its decline may not be as rapid in the forested parts of India, as it mainly feeds on wild ungulates (which are not contaminated with diclofenac).

The Red-headed Vulture previously had less exposure to the toxin owing to competitive exclusion from carcasses by Gyps vultures (Cuthbert *et al.* 2006). With the present decline in Gyps vultures, the Red-headed Vulture now has more access to carcasses, particularly the softer visceral organs such as liver and kidneys, which have the highest concentration of diclofenac (Taggart *et al.* 2006).

Poisoned carcasses placed by villagers for revenge killing of large predators such as tigers and leopards also cause accidental poisoning of vultures. With its

SANJAY KUMAR

Recent scientific studies indicate that the Red-headed Vulture is also impacted by the use of the killer drug diclofenac

already low numbers and slow growth, any further abnormal adult mortality brings the Red-headed Vulture closer to extinction.

Conservation measures underway: The Red-headed Vulture is included in Schedule IV of the Indian Wildlife (Protection) Act, 1972. It is also listed in CITES Appendix II and CMS Appendix II. BirdLife International (2013) and IUCN have listed it as Critically Endangered. Studies on its breeding biology were conducted by BNHS at Keoladeo National Park in the 1990s. Nation-wide survey of all species of vultures is being conducted by BNHS, RSPB, and IBCN. Conservation breeding efforts have not been made as for the Critically Endangered Gyps vultures, and these are urgently needed. Recently, two Red-headed Vulture have been deployed PTTs which are giving valuable data on their movement (Vibhu Prakash *pers. comm* 2014).

RECOMMENDATIONS

(1) Survey Uttar Pradesh extensively to identify the location and number of remaining individuals.

(2) Support the ban on the veterinary use of diclofenac, ketoprofen, and other similar drugs. Promote the immediate adoption of meloxicam as an alternative to diclofenac and ketoprofen.

(3) Test other NSAIDs to identify additional safe alternative drugs to diclofenac and also other toxic ones.

(4) Initiate public awareness and public support programmes along with the programmes for other vulture species.

(5) Conduct ecological and behavioural studies.

(6) Study dispersal and movement through PTTs.

Bengal Florican

Houbaropsis bengalensis (Gmelin 1789)

DHRITIMAN MUKHERJEE

Tall wet grasslands of the Terai in Uttar Pradesh are some of the most important remaining habitats of the Bengal Florican. IUCN and BirdLife International (2001, 2013) list the Bengal Florican in the Critically Endangered category as it has a very small, declining population, a trend that has recently become extremely rapid and is predicted to continue in the near future, largely as a result of widespread and ongoing conversion of its grassland habitat to agriculture.

Field Characters: The Bengal Florican is the size of a domestic hen (height 66–68 cm), with longish legs. The male has mostly black head, neck, breast, and underparts, with back dark brown, heavily mottled and vermiculated, with bold black arrowhead marks. Elongated plumes overhang the breast, more visible before and during display. Wings are largely white, with black primaries. Female and juvenile male with overall rufous-buff, sandy buff, and brown back, mottled with bold black arrowhead marks. No glistening white on wings, instead they have buffish-white wing coverts. Female also has two brown bars on the wing, visible in flight.

There are two subspecies: *Houbaropsis bengalensis bengalensis* found in the Indian subcontinent, and *H. b. blandini* in Cambodia and Vietnam. The subspecies *H. b. blandini* has overall richer plumage tones and the male has shorter ornamental feathers on the head and neck (Johnsgard 1991).

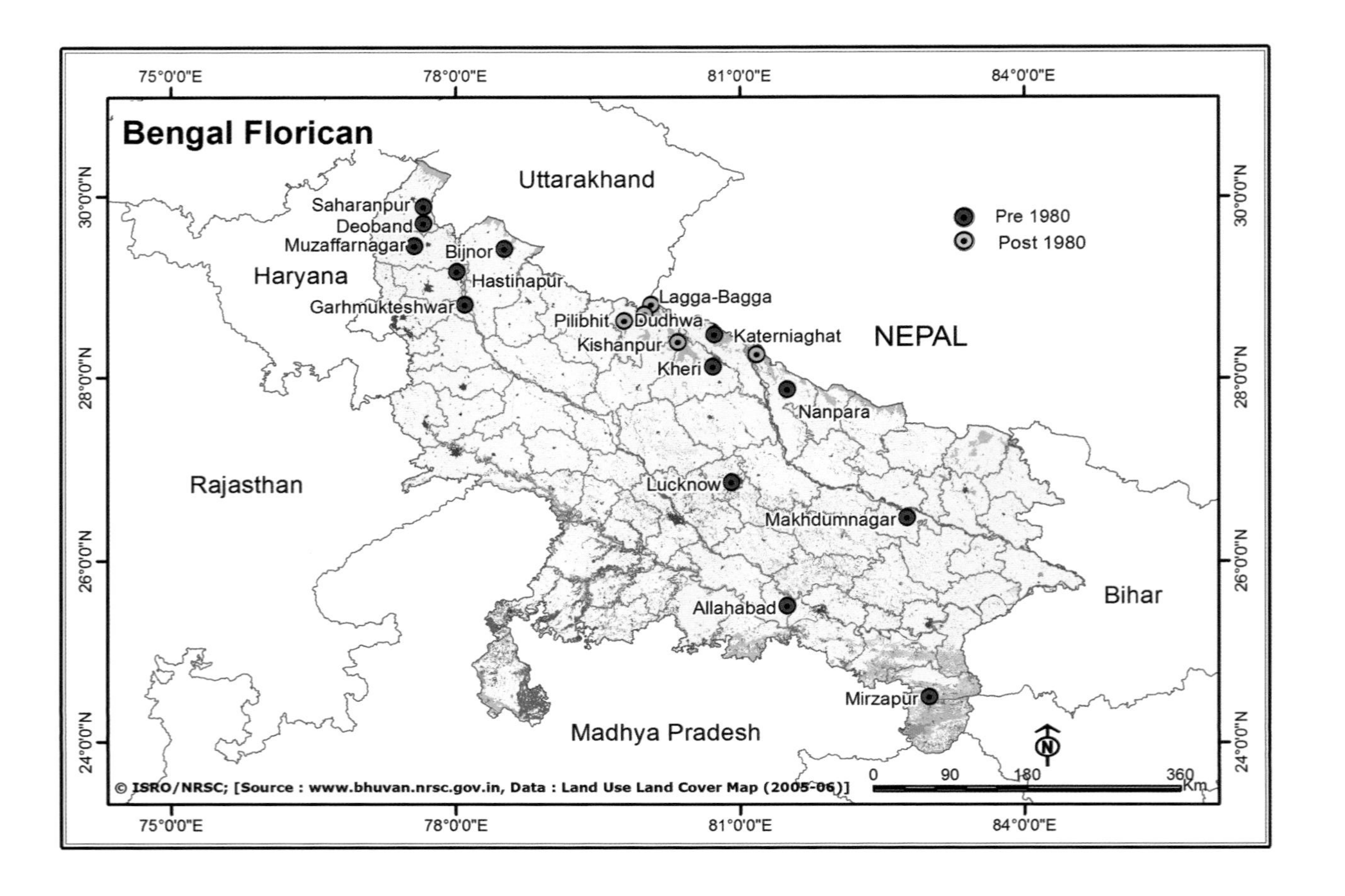
Bengal Florican
Pre 1980
Post 1980
Uttarakhand
NEPAL
Haryana
Rajasthan
Madhya Pradesh
Bihar
Saharanpur
Deoband
Muzaffarnagar
Bijnor
Hastinapur
Garhmukteshwar
Lagga-Bagga
Pilibhit
Dudhwa
Kishanpur
Katerniaghat
Kheri
Nanpara
Lucknow
Makhdumnagar
Allahabad
Mirzapur
0
90
180
360
Km
75°0'0"E
78°0'0"E
81°0'0"E
84°0'0"E
30°0'0"N
28°0'0"N
26°0'0"N
24°0'0"N
© ISRO/NRSC; [Source : www.bhuvan.nrsc.gov.in, Data : Land Use Land Cover Map (2005-06)]

Distribution: The Bengal Florican occurs in India and Nepal and historically in Bangladesh (now extinct). In the Indian subcontinent, it is found or used to be found from the Terai of Uttarakhand, Uttar Pradesh, Bihar, and Nepal to West Bengal, Assam, and lower Arunachal Pradesh. It may still occur occasionally in the extant grasslands of Uttarakhand, but its main population is found in Uttar Pradesh (North Pilibhit, Dudhwa National Park, Kishanpur, and occasional birds in Lagga-Bagga and Katerniaghat). In Nepal, it is found in Sukla Phanta (perhaps some birds from there were seen in Lagga-Bagga), Bardia NP, and Royal Chitwan NP. Some birds from Chitwan could move to Sohagi Barwa Wildlife Sanctuary in Uttar Pradesh and Valmiki Tiger Reserve in Bihar, although there is no confirmed record till now.

Till the late 1980s, in West Bengal it was seen in the private grassland of Sahabad-Sayedabad and Chapra Tea Estates (Rahmani *et al*. 1990), and till the late 1990s in the grasslands of Jaldapara WLS, but there is no recent record (G. Maheswaran *pers. comm*. 2011). According to Sumit Sen (*pers. comm*. 2010), a pair was observed by Bhaskar Das in the Buxa area in 2005.

In Assam, the grasslands of the Brahmaputra valley have always been the stronghold of the Bengal Florican in India. It was widely and extensively distributed from the westernmost part of Assam (Dhubri and Kokrajhar districts) to the easternmost extent of the state up to Sadiya, Dibrugarh, and North Lakhimpur districts. Now it is found in isolated, extant grasslands of Manas TR, Kaziranga, Orang, Dibru-Saikhowa, and Burhachapori, but unfortunately not in Pobitora. Records up to 1990 were given by Rahmani *et al*. (1990), and later up to 2000 by BirdLife International (2001). Recently, Rahmani (2012) has given records up to 2012.

In Uttar Pradesh, it is definitely reported only from Dudhwa NP, Kishanpur WLS, and North Pilibhit Reserve Forest, with earlier reports from Katerniaghat WLS.

Ecology: In India, the ecology and behaviour of the Bengal Florican have been studied in Manas in Assam (Narayan & Rosalind 1990, Narayan 1992) and Dudhwa in Uttar Pradesh (Sankaran & Rahmani 1990; Sankaran 1991, 1996).

Although largely cursorial, the Bengal Florican can fly very well. It is normally solitary in the breeding season, but up to six males may come together for a short period lasting several minutes, and often two females are seen in the same patch of grassland. Most of the adult males become territorial in the breeding season, while a few remain non-territorial, probably due to lack of suitable grasslands. This bird appears to favour relatively open short grasslands (0.5–1.0 m tall) sometimes with patches of tall grass and scattered bushes (Inskipp & Inskipp 1983, Narayan & Rosalind 1990), usually in lowlands below 300 m. The major grass species in these habitats are *Imperata cylindrica*, *Saccharum bengalense*, *Phragmites karka*, *Vetiveria zizanioides*, and *Desmostachya bipinnata*, with or without scattered small trees. Shorter grasses appear to be

DHRITIMAN MUKHERJEE

favoured while foraging or displaying (Sankaran 1996). However, birds seek shelter in tall grass during the heat of the day, and females, which are difficult to see, probably spend much of their time in tall grass, together with males outside the breeding season. Bengal Florican generally avoids large, dense stands of tall grass and is seen in shorter grassland dominated by *Imperata cylindrica*, interspersed with patches of taller grassland (Narayan & Rosalind 1990). Similar behaviour was noticed in Nepal by Peet (1997) and in Cambodia by Davidson (2004).

During the breeding season, males go back to their traditional display territories. Some territories have been occupied over the last 30 years of monitoring (e.g., Sonaripur grassland in Dudhwa, Kohora in Kaziranga). Good territories are at a premium and adult males fight for them. Flight display of the Bengal Florican was first properly described by Sankaran (1991) and Narayan (1992). The flight display usually takes place in an open patch of the male's territory. Once the male is aroused, it fluffs up the head, neck, and breast feathers. Just before taking the jump, it inflates the breast pouch even further, draws the head further back, and lowers the body by partly bending the legs. The bird springs forward at an angle of about 45°. A loud and rapid wing-flapping sound is made while ascending and on reaching a peak 3–4 m high, where the flapping stops and the wings are opened, displaying the glistening white wing feathers vividly against the jet black body. Then it delivers its sharp, whistle-like *chip-chip* call. It glides down a metre or two, moving forward on open wings, with the pouch drooping under the breast and the head thrown back. Just 1–2 m above ground, it begins to flap its wings again and moves forward, gaining lost height. On reaching the apogee, it stops flapping its wings and floats down more or less vertically with partly open wings, drooping pouch, dangling and even paddling legs. During the display flight, it covers anything between 20–40 m ground and takes 6–8.5 seconds from take-off to landing. It calls four to seven times while in the air.

For details of ecology and behaviour, see Sankaran (1991), Narayan (1992), and BirdLife International (2001).

Threats: Destruction of its grassland habitat is the biggest threat to the Bengal Florican. Some of the other threats are poaching, overgrazing, burning of grassland during the breeding season, and overall disturbance in the breeding season. We do not know what happens when it moves out of the breeding grounds in the non-breeding season, although it occurs in some very well-protected areas of UP (e.g., Dudhwa, Kishanpur, North Pilibhit). Even inside these PAs, it gets only incidental protection and no attempt has been made to initiate Bengal Florican-specific habitat protection. In most of the PAs, grassland is burnt or cut even during the peak breeding season of the Bengal Florican. Late burning of grassland, while the Bengal Florican is breeding, may be one of the major threats to nesting females, eggs, and chicks. There are also site-specific threats for which we have given recommendations. In places like Katerniaghat which

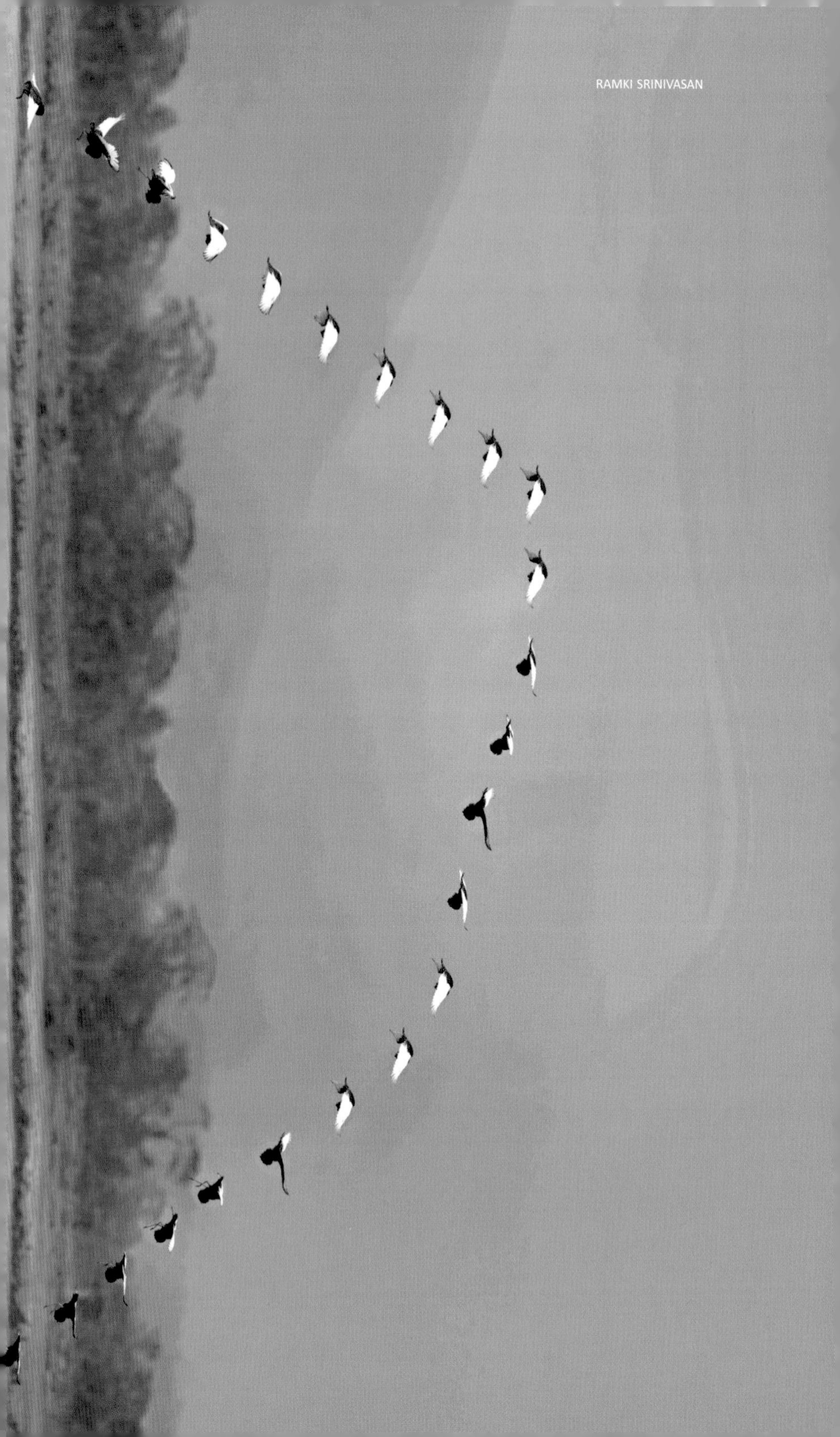
RAMKI SRINIVASAN

DHRITIMAN MUKHERJEE

le Bengal Florican has spectacular ground (above) and aerial (below) courtship displays to attract females and to advertise its territory

DHRITIMAN MUKHERJEE

If properly implemented, the Bengal Florican Conservation Plan can benefit local people by providing them sustainably harvested grass. However, the grass must be harvested by January and a no-disturbance policy should be applied from February to June to allow undisturbed breeding

have potential grassland habitats which the Bengal Florican can recolonise, a large number of villagers are allowed to pluck *Grewia* (Falsa) fruit in the florican breeding season. Thatch collection and livestock grazing are perennial problems in Katerniaghat and North Pilibhit Forest. Invasive species such as *Mimosa*, *Eupatorium*, *Mikenia*, *Leea*, and *Dillenia pentagyna* have also contributed to habitat degradation. Grasslands are still being afforested by the Forest Department (e.g., Compartment No. 30 of Sohagi Barwa WLS in Uttar Pradesh).

Conservation measures underway: It is included in Schedule I of the Indian Wildlife (Protection) Act, 1972 and in CITES Appendix I and II. Several populations occur within protected areas, the most important being Chitwan NP, Bardia NP, and Sukla Phanta Wildlife Reserve in Nepal; Dudhwa, Pilibhit, and Kishanpur in Uttar Pradesh; Kaziranga, Manas, Orang in Assam; and D'Ering in Arunachal Pradesh. In Southeast Asia, a tiny population may still remain at Tram Chim National Park, Vietnam, and another at Ang Trapeang Thmor Sarus Crane Conservation Area, Cambodia.

A species recovery plan has been prepared by BNHS, WII and WWF-India in collaboration with state governments and the Ministry of Environment and Forests (MoEF), Government of India.

rt grassland with scattered trees is ideal habitat for Bengal Florican. Only tracking studies (PTT) will ll us the habitat requirement of female Bengal Florican, particularly for nesting and chick rearing

A project to study the movement and dispersal of the Bengal Florican was initiated in Dudhwa and other areas in May 2013, funded by MoEF, Government of India, and partially funded by Darwin Initiative for the Survival of Species, UK.

Recommendations: BirdLife International (2001, 2013) has given recommendations for all the range countries of Bengal Florican, and Rahmani (2012) has given India-specific recommendations. Here we give specific recommendations for its protection in UP.

General Recommendations

(1) Project Bustards: The Government of India should start Project Bustards which should protect Bengal Florican and its alluvial grassland habitat.

(2) Research and Monitoring: Long-term research and monitoring of the Bengal Florican should be taken up on an urgent basis. At present, there is some information about the status of Bengal Florican in a few protected areas but not in its entire distribution range. In addition, very limited research has been undertaken on its habitat and breeding biology. There is no information at all about its sex ratio, life cycle, food preferences, and food availability in its habitats. Therefore, to gather this data more research should be taken up. Species Recovery Plan should be fully implemented as a part of Project Bustards in all the range states (Uttar Pradesh, Uttarakhand, Assam, Arunachal Pradesh, and West Bengal).

Intensive and repeated grazing by livestock is the biggest threat to the shy Bengal Florican. Livestock grazing in known and potential florican areas should be stopped immediately under the Bengal Florican Recovery Plan of the MoEF

(3) Strengthening protection along Protected Area boundaries: Many of the Bengal Florican grasslands are situated along Protected Area boundaries, and such established Bengal Florican territories are under threat from anthropogenic disturbances. Therefore, to safeguard these habitats, protection should be strengthened along the Protected Area boundaries.

(4) Control grazing, collection of food plants and thatch: Bengal Florican is a sensitive species, and a grassland habitat specialist, and cannot adapt to disturbances in its habitat. Hence, intensive anthropogenic disturbances like grazing by livestock, collection of food plants and thatch material will have drastic effects on its territories. In fact, already there is evidence of disappearance of Bengal Florican territories from highly disturbed grassland areas. Therefore, to ensure protection to these territories and grassland habitats, grazing, collection of food plants, and thatch collection should be completely stopped.

(5) Control repeated grass burning: Repeated grass burning results in degradation of habitat quality and loss of cover for the species. Hence in order to avoid habitat degradation, burning should be done only once a year and should be completed in Bengal Florican habitats before January 15.

(6) Eradication of invasive species: In recent years, spread of invasive species in grasslands is a major conservation issue in Protected Areas. Some of the most important grassland habitats are affected by invasive species such as *Eupatorium*, *Mimosa*, and *Mikenia*. To protect these habitats, eradication of invasive species is the most effective method.

(7) Information on non-breeding habitats and behaviour of Bengal Florican should be gathered: Till date very little research has been undertaken on non-

breeding habitats of Bengal Florican and almost no information is available on its non-breeding behaviour. Therefore, more research should be undertaken to gather information in these areas.

(8) Preparation of habitat reclamation plan: Potential grassland habitat in new areas should be identified and restored for species like Bengal Florican. There should be a proper action plan to protect these newly identified grasslands, including habitat management. Moreover, at least an attempt should be made to reclaim the Bengal Florican habitat where encroachment is limited to cultivation, and till now no human habitations or villages have come up. An effective habitat reclamation plan should be prepared.

(9) Increase in the level of awareness among target stakeholders: Though Bengal Florican is a Critically Endangered species, even stakeholders like decision makers, PA managers, PA frontline staff, and people living in the fringe areas around protected areas do not know enough about the importance of this bird. To enhance awareness on the species and its habitat, there should be an organized education programme for target stakeholders.

Specific Recommendations

(1) The derelict Seed Farm in Katerniaghat WLS which was recently taken over by the Forest Department should be developed as a grassland. There should be a total ban on tree plantation in this area and the existing Eucalyptus trees should be removed.

(2) In tall grassland areas, small patches of 2–4 ha should be cleared by periodic manual cutting of grass to create Bengal Florican male display territories. Care should be taken that this should not be attempted where Bengal Florican already has territories.

(3) The grasslands of North Pilibhit forests should be strictly maintained as grasslands with no plantation whatsoever. The existing plantations which have matured should be removed and grasslands should be restored.

(4) Lagga-Bagga is an ecological extension of Sukla Phanta of Nepal. The grassland should be maintained and improved, with no tree plantation.

(5) In Katerniaghat WLS, collection of *Grewia* sp. (Falsa) fruit by villagers should be totally prohibited as this activity creates huge disturbance.

(6) Cattle grazing in Katerniaghat which is rampant at present should be strictly controlled. This looks impossible at present but the UP Forest Department has set many good examples of wildlife management. For instance, cattle camps of Sonaripur Range were successfully removed nearly 30 years ago.

Baer's Pochard

Aythya baeri (Radde 1863)

DHRITIMAN MUKHERJEE

Formerly classified as Vulnerable (BirdLife International 2001), Baer's Pochard has been uplisted to Endangered owing to an apparent acceleration in its decline, measured by its numbers on the wintering grounds. It is now absent or occurs in greatly reduced numbers over much of its former wintering range and is common nowhere (BirdLife International 2013). It is a marginal species in Uttar Pradesh, with only one confirmed record from Okhla Bird Sanctuary.

Baer's Pochard breeds in eastern Russia, northeast China, and possibly in Mongolia and North Korea, and has been recorded on passage or in winter (or as a vagrant) in Mongolia, Japan, North Korea, South Korea, mainland China, Hong Kong, Taiwan, Pakistan, India, Nepal, Bhutan, Bangladesh, Myanmar, Thailand, Vietnam, and the Philippines (BirdLife International 2001). The main wintering areas appear to be in eastern and southern China, northeast India, Bangladesh, Thailand, and Myanmar.

In India, it was an uncommon and erratic winter visitor to Assam and West Bengal, and to Manipur fairly regularly. It has been recorded in other states as well (Rahmani & Islam 2008). Possibly it is less rare in India than recorded, being overlooked or confused with the Ferruginous Duck. Both the species are found together and the females are superficially alike (Ali & Ripley 1987, Grimmett *et al*. 1998). Rahmani (2012) has given records from India.

Javed & Rahmani (1998) during their studies on birds in Dudhwa NP from 1991 to 1994 did not record this species. Manoj Sharma (*pers. comm*. 2013), who has been birding for more than a decade, has not recorded this species anywhere in Uttar Pradesh.

Harris (2001) mentioned a pair seen on January 11, 2001 by W. Harvey at **Okhla** Bird Sanctuary which lies across Uttar Pradesh and Delhi. As it is marginal to UP, we are not describing it in detail.

RECOMMENDATIONS

As it is a scarce migratory species in India, greater protection of wetlands, prevention of poaching and poisoning of waterfowl would greatly help the species. However, it is more important to protect its nesting areas by establishing more protected areas in its breeding grounds.

Egyptian Vulture

Neophron percnopterus (Linnaeus 1758)

The Egyptian Vulture is perhaps the most widespread vulture of the Old World, with isolated resident populations in Cape Verde and Canary Islands off the northwest coast of Africa, north Africa, Ethiopia, and east Africa, isolated populations in Angola and Namibia, southern Europe, the Mediterranean, the Middle East, Central Asia to India and Nepal. In its wide range, it is declining rapidly, therefore BirdLife International (2013) has listed it as Endangered. It is a long-lived and slow breeding bird with very few predators on the adult, therefore any decrease in breeding or increase in adult mortality, as seen in southern Europe (>50% over the last three generations, i.e., 42 years) could spell doom for this species. India, where good populations used to be present 20 years ago, has also seen a sharp decline. BirdLife International (2013) estimates its total world population between 20,000 and 67,000 mature individuals.

Field Characters: The Egyptian Vulture is a small kite-like vulture with naked head and short, all-feathered neck. Adult dirty white with black flight feathers, while juvenile is dark with pale vent and tail. The face is bare, yellow in adult and brown in juvenile.

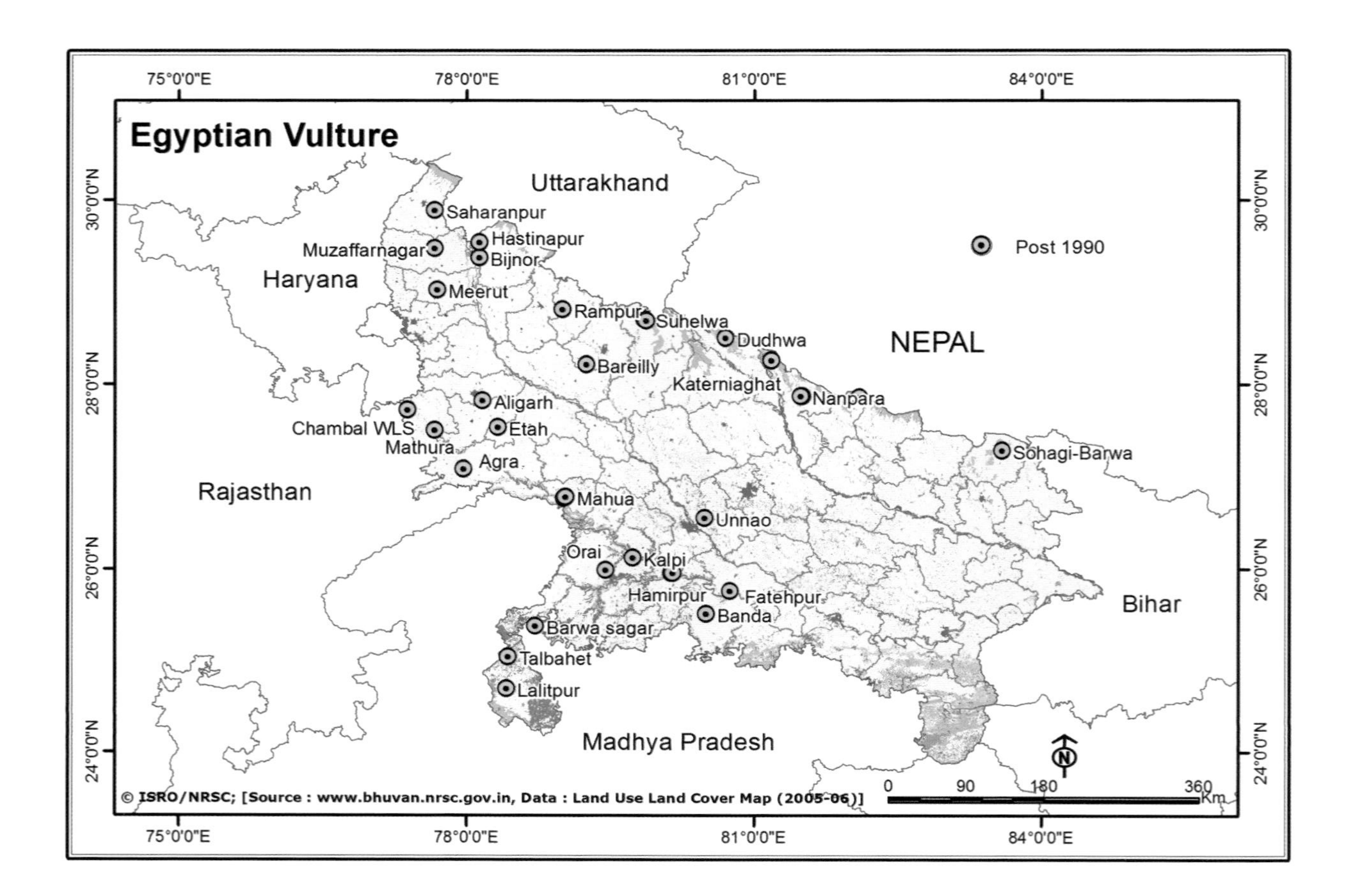

Egyptian Vulture
Post 1990
NEPAL
Uttarakhand
Haryana
Rajasthan
Madhya Pradesh
Bihar
Saharanpur
Hastinapur
Bijnor
Muzaffarnagar
Meerut
Rampur
Bareilly
Suhelwa
Dudhwa
Katerniaghat
Nanpara
Sohagi-Barwa
Aligarh
Etah
Agra
Mathura
Chambal WLS
Mahua
Unnao
Orai
Kalpi
Hamirpur
Fatehpur
Banda
Barwa sagar
Talbahet
Lalitpur
75°0'0"E
78°0'0"E
81°0'0"E
84°0'0"E
30°0'0"N
28°0'0"N
26°0'0"N
24°0'0"N
0 90 180 360 Km
© ISRO/NRSC; [Source : www.bhuvan.nrsc.gov.in, Data : Land Use Land Cover Map (2005-06)]

While adult Egyptian Vulture is dirty white to bright white (breeding) in colour, the juvenile is dark, with pale vent and tail

Distribution: The Egyptian Vulture has a very wide range in Africa, southern Europe, the whole of the Middle East, Iran, Afghanistan, Pakistan, India, and Nepal. For details of population figures, see BirdLife International (2013). It is found all over India, from the plains to *c.* 2,500 msl, sometimes very close to human habitation. It is still widespread in India and frequently seen in Uttar Pradesh, Uttarakhand, Rajasthan, Gujarat, Madhya Pradesh, Chhattisgarh, Maharashtra, and decreasingly so in south India. Rahmani (2012) has given some recent records.

Records from Uttar Pradesh: It is not a common bird in Uttar Pradesh anymore. Earlier, it was fairly abundant in areas of central Uttar Pradesh, Bundelkhand and Brij (Agra, Mathura, Etawah, Etah districts), and parts of the Terai. Even now some stray records are available from protected areas, for example, sighting of two birds in **Katerniaghat** WLS in January 2008 (Amit Mishra *pers. comm.* 2010) and five birds in Tulsipur Range of **Sohelwa** WLS in 2010, feeding on a carcass with other species of vultures. In most of our sightings, it was usually found near or at garbage dumps on the outskirts of a town.

Nikhil Shinde and Rajat Bhargava of BNHS during their studies on birds in Soheldev and Sohagi Barwa WLS from November 2013 to March 2014 have seen Egyptian Vulture on the following sites: four birds on December 28, 2013 in Poorvi Sohelwa Range, **Suhelwa** Sanctuary; two on February 7, 2014 in South Chowk Range, **Sohagi Barwa** Sanctuary; 13 on December 12, 2013 outside **Nanpara** town near carcass dumps; and two on March 30, 2014 in **Pilibhit** Reserve Forest.

The Egyptian Vulture is generally found near slaughterhouses and carcass dumps, as seen above near Nanpara

We noticed more than 50 birds on the left bank of Ken river in **Banda** district, where these birds were feasting on a carcass amidst all sorts of litter. Another encounter with this bird was outside **Fatehpur** district (between Kanpur and Allahabad) where a flock of more than 50 birds was seen in the company of Pariah Kites *Milvus migrans migrans*, feeding on a dead animal. Some sporadic records are from **Unnao**, a town known for its leather products, mainly footwear. Besides, a large meat processing unit is functional here. This may be the main reason for regular sightings of these vultures from this area.

In western Uttar Pradesh, this vulture is the most commonly seen vulture species, nevertheless its numbers are extremely low compared to earlier times. During the last two years of birding records during travelling, this vulture species was encountered (Rajat Bhargava *pers. obs.* 2012–13), among which notable records are from the following areas (mainly around abattoirs): *c.* 15 birds including juveniles near human habitation *c.* 10 km on the outskirts of **Saharanpur** towards Yamunanagar; *c.* 10 birds including juveniles on the outskirts of **Muzaffarnagar** towards Khatuli; six birds (including 2 juveniles) on the outskirts of **Bijnor**; *c.* 20 birds including 30% juveniles and subadult along a 30 km stretch of the **Meerut-Hapur** road and Hapur-Pilakhwa road towards **Ghaziabad/ Ghazipur**; 8–10 birds with juveniles on the outskirts of **Rampur** towards Bilaspur; 10 or more birds along the Meerut bypass near **Baraut**; and lastly 6–8 birds along the Meerut-Mawana-**Hastinapur** road. It is regularly sighted in small numbers in the Terai. For example, 13 were seen on **Nanpara** bypass near a slaughterhouse. Interestingly, only four were adult, the rest juvenile or subadult, indicating successful breeding in the area. Kaajal Dasgupta (*pers. comm.* 2013) found 8–10 throughout the year near a carcass dump in **Bareilly**, and up to 6 birds in **Pilibhit**. Other districts where they are frequently seen are **Aligarh** and **Agra**, mostly near garbage dumps.

Ecology: The Egyptian Vulture can be seen sauntering around villages and nomad camps, looking for carrion, offal, garbage, and human excrement. It opportunistically picks up crickets, frogs, and alates of emerging termites. It has a narrow long beak, which helps it in tearing off small pieces of meat through narrow spaces between bones that large-beaked vultures cannot reach. It is usually solitary or found in pairs with juveniles, but on good feeding sites (e.g., Jorbeer carcass dump, Bikaner, Rajasthan), 1000–2000 are seen in winter (Vibhu Prakash *pers. comm.* 2010). It roosts singly or in small groups, generally on tall trees, but electric pylons are frequently used where tall trees are absent (e.g., Rann of Kutch, Gujarat). Although it is mostly resident and seen around its usual haunts throughout the year, the northern populations undertake short to long distance migration as conditions become unsuitable during winter. It feeds on dead animals but can also kill stranded fish and turtles, and perhaps small prey. It mainly nests on cliffs, rocky outcrops, ledges of occupied buildings, abandoned forts and ruins, but occasionally on tall trees where its preferred nesting habitat is not available. A single egg is laid and both parents share incubation and chick-rearing.

Threats: In its vast distributional range, threats vary from country to country and region to region. In **India/Uttar Pradesh**, the main threat could be poisoning by feeding on cattle carcasses contaminated with diclofenac, as has been seen in

ASAD R. RAHMANI

In the near absence of Gyps vultures, and scarcity of Egyptian and Red-headed Vultures, rotting carcasses are now mainly attended by dogs and crows, compounding an ecological nightmare

Gyps species of vultures. Other NSAIDs used in livestock could be killing these vultures also. Earlier, when Gyps species of vultures were in abundance, they would not allow the Egyptian Vulture to feed on the internal organs of a carcass such as lungs and liver, but now with the near total disappearance of Gyps vultures from the Indian subcontinent, the Egyptian Vulture has a greater chance to feed on such internal organs that contain more diclofenac than the muscles and tendons on which it fed earlier. Thus the risk increases. In **India**, the annual rate of population decline was 35% during 2000—2003 and the population in 2003 was estimated to be 20% of that in the early 1990s (Cuthbert *et al*. 2006).

As bird trapping and trade is quite common in UP, a minor threat reported during TRAFFIC India studies on bird trade was trapping of this vulture for zoos (Ahmed 2012).

Conservation measures underway: It is listed in Schedule IV of the Indian Wildlife (Protection) Act, 1972. The veterinary use of diclofenac has been totally banned by Government of India since 2006. Regular surveys by BNHS, funded by RSPB, are ongoing in India.

RECOMMENDATIONS

(1) Total ban on veterinary use of diclofenac should be implemented in the state/ country. Human-use diclofenac should not be sold for veterinary use.

(2) Study the impact of other NSAIDs on Egyptian Vulture.

(3) Conduct surveys in Uttar Pradesh on a regular basis to study the population trend.

(4) Start an environmental education campaign in rural areas about the importance of vultures.

(5) Conduct ecological and behavioural studies on this species.

(6) Study its movement through satellite tracking to map its home range in breeding and non-breeding seasons, and also study the dispersal of juveniles.

Lesser Florican

Sypheotides indicus (Miller 1782)

ASHOK CHAUDHARY

The Lesser Florican, one of the threatened birds of India mainly due to destruction of its grassland habitats and hunting, qualifies as Endangered because it has a very small, declining population. The rate of decline is predicted to increase in the near future, as pressure on the remaining grasslands intensifies (BirdLife International 2013). It is a marginal species to Uttar Pradesh with only a few recent records

Distribution: The Lesser Florican is endemic to the Indian subcontinent. It is an irregular local migrant, and also nomadic in the rainy season (SW monsoon). It was once widespread and common, but now breeds only in a few areas in Gujarat, southeast Rajasthan, northwest Maharashtra, and western Madhya Pradesh. There is some dispersal to southeast India in the non-breeding season. It was once common in the Terai of Nepal, but now it is rarely seen in the area. It has also been sighted in Pakistan and is a vagrant in Bangladesh. For a review of its former distribution, see Sankaran *et al*. (1992) and BirdLife International (2001).

In **Uttar Pradesh**, it is reported from **Dudhwa** National Park, and there is one record from **Mainpuri** district (Sundar 2006). Ravi Singh (*pers. comm*. 2012) has seen it in Amausi Airport near **Lucknow**.

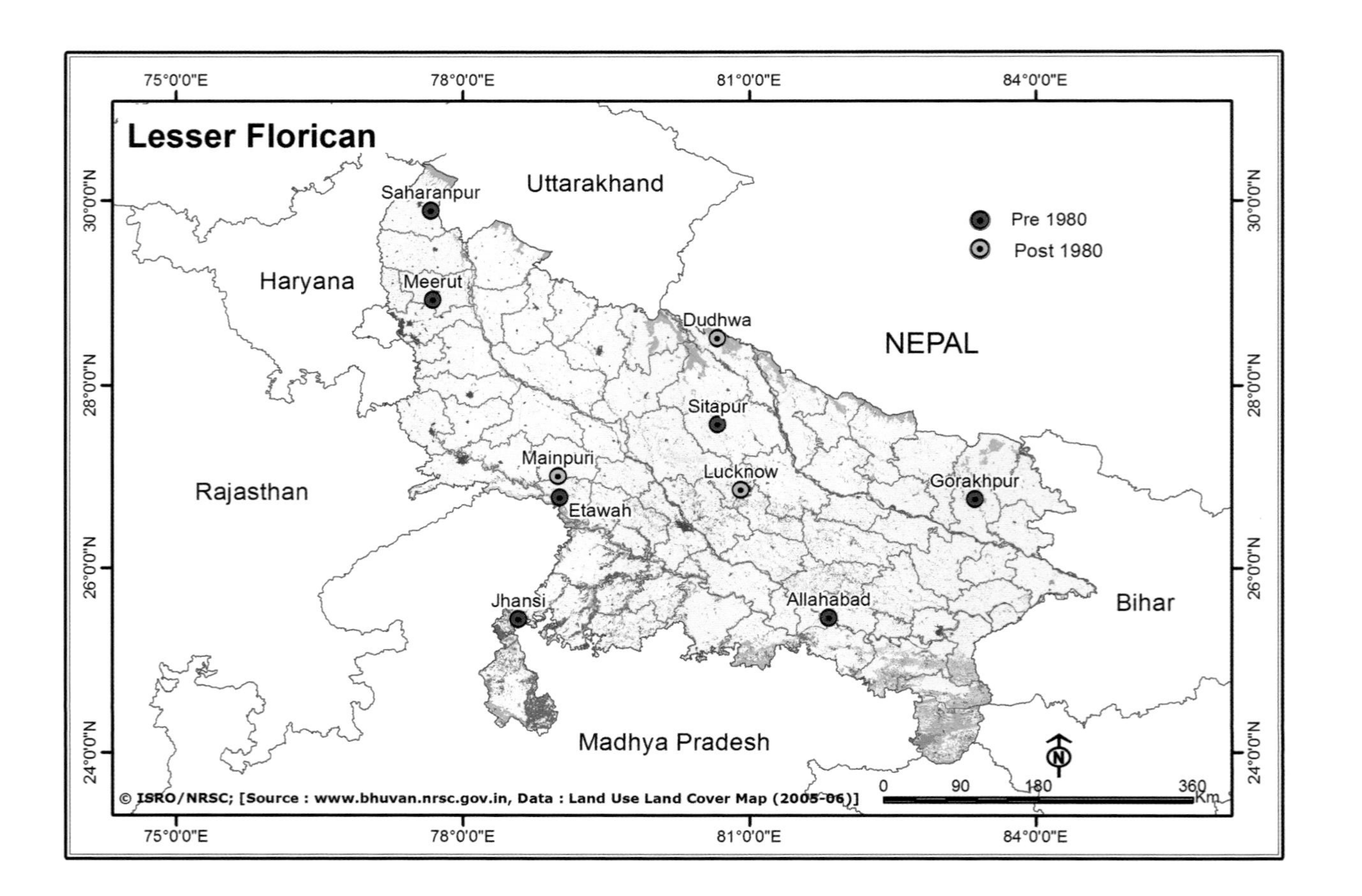
Lesser Florican
Pre 1980
Post 1980
Uttarakhand
Haryana
Rajasthan
NEPAL
Madhya Pradesh
Bihar
Saharanpur
Meerut
Dudhwa
Sitapur
Mainpuri
Etawah
Lucknow
Gorakhpur
Jhansi
Allahabad
75°0'0"E
78°0'0"E
81°0'0"E
84°0'0"E
30°0'0"N
28°0'0"N
26°0'0"N
24°0'0"N
0 90 180 360 Km
© ISRO/NRSC; [Source : www.bhuvan.nrsc.gov.in, Data : Land Use Land Cover Map (2005-06)]

In the 1980s, Lesser Florican was sighted in Navalkhad grassland of Dudhwa National Park

Threats: Earlier, hunting was the main threat as this florican was considered a gamebird, but now destruction and deterioration of its grassland habitat is the main threat, although it is still poached by tribals in many areas (e.g., Jhabua district, northwest Madhya Pradesh). In some areas, invasive species such as *Prosopis chilensis* threaten habitat quality (e.g., Lala Bustard Sanctuary in Gujarat). Over the last two decades, erratic monsoon rains have caused significant population fluctuations.

Conservation measures underway: It is included in Schedule I of the Indian Wildlife (Protection) Act, 1972 and listed in CITES Appendix II. Two sanctuaries were declared in Madhya Pradesh (Sailana in Ratlam and Sardarpur in Jhabua districts) for the Lesser Florican. Good florican habitat is present in Lala and Naliya grasslands in Kutch, and Velavadar NP in Bhavnagar district, Gujarat. The Rajasthan Forest Department has deployed watchmen and cattle guards in Sonkhaliya area in Ajmer district to protect both Great Indian Bustard and Lesser Florican. There are many private initiatives also, which are quite effective, albeit on a small scale. The MoEF, with the help of BNHS, WWF, WII and state forest departments, has prepared the Lesser Florican Recovery Programme (Dutta *et al*. 2013 and MoEF website).

RECOMMENDATIONS

As it is marginal to Uttar Pradesh, we are not giving detailed recommendations which can be seen in Rahmani (2012). Site-specific recommendations have been made in the Bustard Species Recovery Plan (Dutta *et al*. 2013).

Black-bellied Tern

Sterna acuticauda Gray 1832

DHRITIMAN MUKHERJEE

According to BirdLife International (2013), the Black-bellied Tern is almost extinct in a large part of its range, but remains locally common throughout the Indian subcontinent. Consequently, overall decline may be moderately rapid, qualifying the species for Endangered status, although monitoring is urgently needed to reassess trends in India. Its total population could be between 10,000 and 25,000.

Field Characters: A small (33 cm) tern with deeply forked tail and deep orange bill. During the breeding season, the adult has black cap, white lores, dark grey breast, and black belly and vent. Non-breeding birds and immature birds have white belly, with streaked crown and black mask. Juveniles are buffy grey above with blackish markings, lacking tail streamers, and the orange bill has a dark tip.

Distribution: The Black-bellied Tern is found on all the major rivers of South and Southeast Asia. It is not reported from Sri Lanka (Kotagama & Ratnavira 2010). In **India**, it is resident in all the major rivers of north, central, and eastern India, becoming uncommon southwards where it is a winter migrant. It is essentially an inland and freshwater tern, not found on the sea coast (Ali & Ripley 1987).

Records from Uttar Pradesh: In **Uttar Pradesh**, it has been specifically reported from **Lakh-Bahosi** Bird Sanctuary, **Narora**, **Patna Bird Sanctuary, River Ramganga** between Kalagarh and Moradabad, **River Gomti**, and in **Jaunpur** and **Barabanki**.

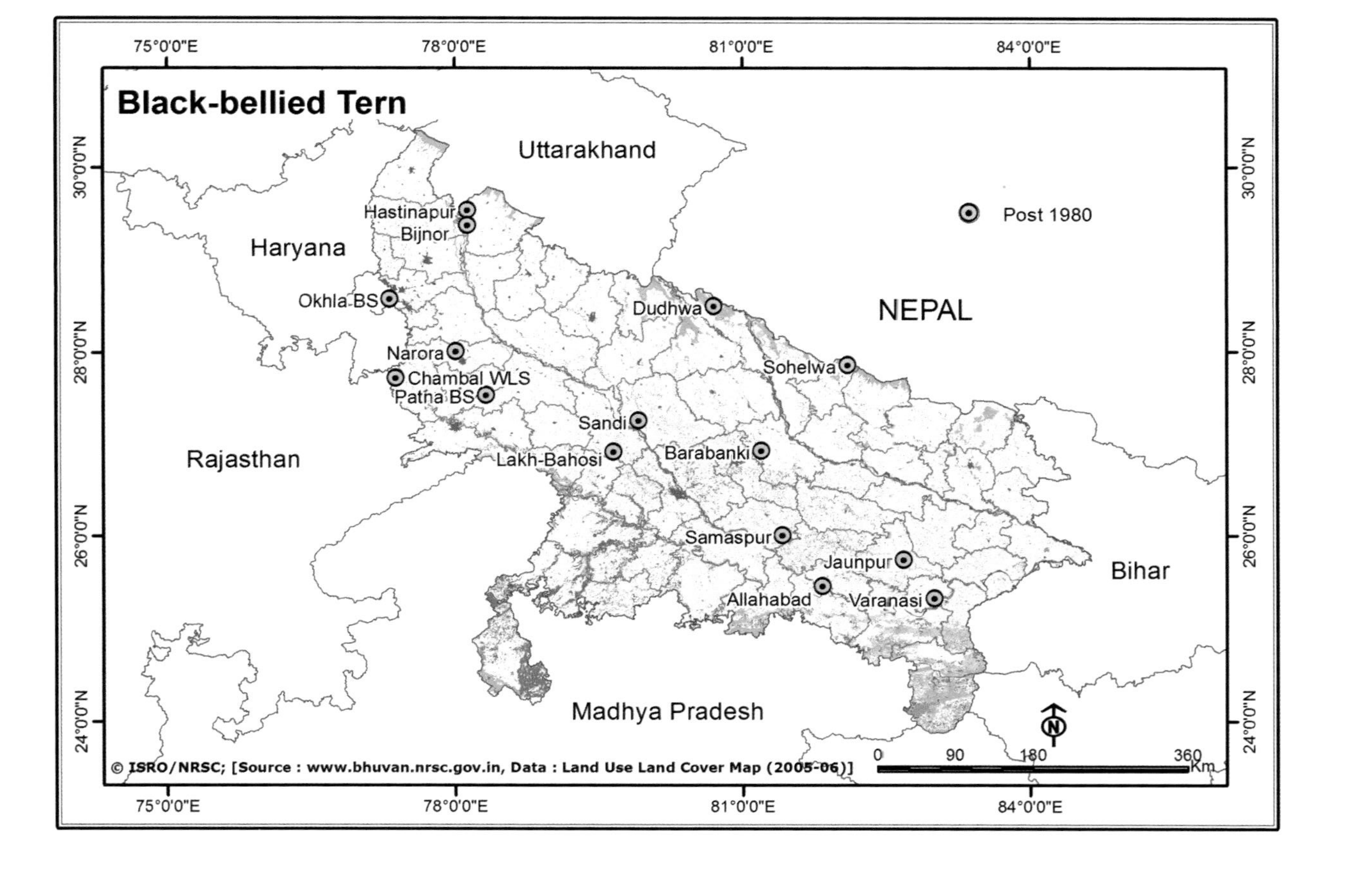
Black-bellied Tern
Post 1980
Uttarakhand
Haryana
NEPAL
Rajasthan
Madhya Pradesh
Bihar
Hastinapur
Bijnor
Okhla BS
Dudhwa
Sohelwa
Narora
Chambal WLS
Patna BS
Sandi
Lakh-Bahosi
Barabanki
Samaspur
Jaunpur
Allahabad
Varanasi
75°0'0"E
78°0'0"E
81°0'0"E
84°0'0"E
30°0'0"N
28°0'0"N
26°0'0"N
24°0'0"N
0 90 180 360 Km
© ISRO/NRSC; [Source : www.bhuvan.nrsc.gov.in, Data : Land Use Land Cover Map (2005-06)]

River banks and sandy islands that are used by River Tern and other species are now mostly occupied by humans, leaving very few undisturbed areas for these birds to breed

It can be easily spotted near the *ghat*s on the banks of the Ganga in **Narora**, **Varanasi**, and **Allahabad**. It is likely to be present in many more areas. Javed & Rahmani (1998) reported it as vagrant in the wetlands of **Dudhwa** NP.

Ganguli (1975) recorded it as resident and fairly common on the River Yamuna (passing through Delhi and UP), but now it is a scarce summer visitor, according to Harris (2001), who saw six birds in September 2001 in **Okhla** Bird Sanctuary. Mike Prince photographed a bird in Bateshwar on the **Yamuna** river in February 2005. Bhargava (2012) has reported it from **Hastinapur** WLS, Meerut and has also recorded two birds at **Bijnor Barrage** in 2010.

R.K. Sharma (*in litt.* 2010) conducted annual wildlife surveys in the **National Chambal Sanctuary** in February-March from Pali to Pachnada, a distance of 435 km. He counted the following numbers of Black-bellied Tern (year given in brackets): 24 (2003), 34 (2004), 46 (2005), 64 (2006), 55 (2007), 59 (2008), 54 (2009), and 58 (2010).

During a waterfowl census in February 2008, six Black-bellied Tern were spotted in **Hakimpur** wetlands within the Narora Atomic Power Plant in Bulandshahr district (Raja Mandal *pers. comm.* 2008). Being a denizen of large rivers, it is not commonly reported from most of the major wetlands, though regular sight records are available from **Sandi** and **Samaspur** Wildlife Sanctuaries. Two birds amidst a group of River Tern were reported from **Samaspur** during March 2008. It has also been reported from **Narora** for three consecutive years: 17 birds in January 2011 (Raja Mandal & P.D. Mishra *pers. comm.* 2012), 25 in February 2012, and 20 in February 2013 from **Narora Barrage** (Amit Mishra *pers. comm.* 2013) over the River Ganga. Sanjay Kumar has also seen it near **Narora** in

January and February 2014. His other records are from **Geruwa** and **Kaudhiala** in Katerniaghat WLS, **Chambal** river (December 2013 and February 2014) and **Gomti** river near Lucknow.

Ecology: It inhabits large rivers, foraging methodically over long stretches of placid waters, and resting on river islands and sandbanks. It feeds mainly on fish, also insects and crustaceans. It is gregarious and hunts in groups. It breeds colonially in summer (April to June) in the north, and February onwards in the south. The nest is a mere scrape on the sand, where two or three eggs are laid. Sometimes the nests are quite far apart, but it also nests colonially with other terns, Indian Skimmer, and pratincoles (del Hoyo *et al.* 1996). Wide nest dispersion is presumably an adaptation to heavy pressure from terrestrial predators, including man. Incubation and fledging periods are unknown. Call is a pleasant *krek-krek*, constantly uttered as it flies about (Baker 1929). Not much is known about its breeding and feeding ecology.

Threats: As human population increases, this bird faces numerous threats, particularly during the breeding period. Most of the large rivers in South Asia are now dammed and their islands heavily cultivated, leaving not many undisturbed areas for it to breed. As a result of dams and utilisation of water (through pumps and pipes) for cultivation or supply to towns and villages for drinking, there is very little water left in rivers in summer, exposing the riverine islands to terrestrial predators. There is not much collection of eggs for food in India, but dogs, cats, and crows destroy whole colonies. Sudden release of water from dams also washes away eggs and chicks. Occasionally they are accidentally caught in nets laid for other waterbirds and opportunistically traded (Ahmed 2002).

Conservation measures underway: Like most wild birds, it is listed in Schedule IV of the Wildlife (Protection) Act, 1972. It is found in a large number of PAs/IBAs.

RECOMMENDATIONS

1) Study its breeding and feeding ecology, with special emphasis on the threats to its breeding areas.

(2) Study its movement through marked birds.

(3) Study the impact of removal or sudden release of water into large rivers on its breeding success.

(4) Identify major breeding areas in large rivers and involve local communities in the protection of nests.

(5) Control the populations of stray dogs, cats, and crows in its breeding areas to increase breeding success.

(6) Strictly prohibit trapping and trade of this species and all other gulls and terns.

Swamp Francolin
Francolinus gularis (Temminck 1815)

SANJAY KUMAR

As the name implies, this is a bird of swamps and damp areas, but it is doubtful if it ever goes into the water. Its specialised habitat, the Terai in north India and Nepal, extends from north Pilibhit district in Uttar Pradesh to West Bengal. It is also found in the Brahmaputra floodplains in Assam. Earlier it was found in Bangladesh, but there are no recent records. Based on the decline of its specialised habitat during the last 40–50 years, which is projected to continue, concurrent with the rapid degradation in habitat quality, and hunting/trapping pressures, BirdLife International (2001, 2013) and IUCN consider it Vulnerable.

Field Characters: The Swamp Francolin, earlier erroneously called Partridge though it is not a true partridge (*Perdix* species), is the largest francolin found in India. It has conspicuously long legs (an adaptation to its swamp habitat) and a densely-patterned body: upperparts brown with rufous-brown bars, lower part brown with long longitudinal white streaks edged with black. Crown and nape brown, supercilium and broad line below the eye buff. Chin and throat rusty red (dark rufous), extending to the upper neck. Undertail-coverts pale rufous, so also the primaries — conspicuous when the bird takes flight. Sexes alike, but the male has a spur on each leg and is also slightly bigger than the female.

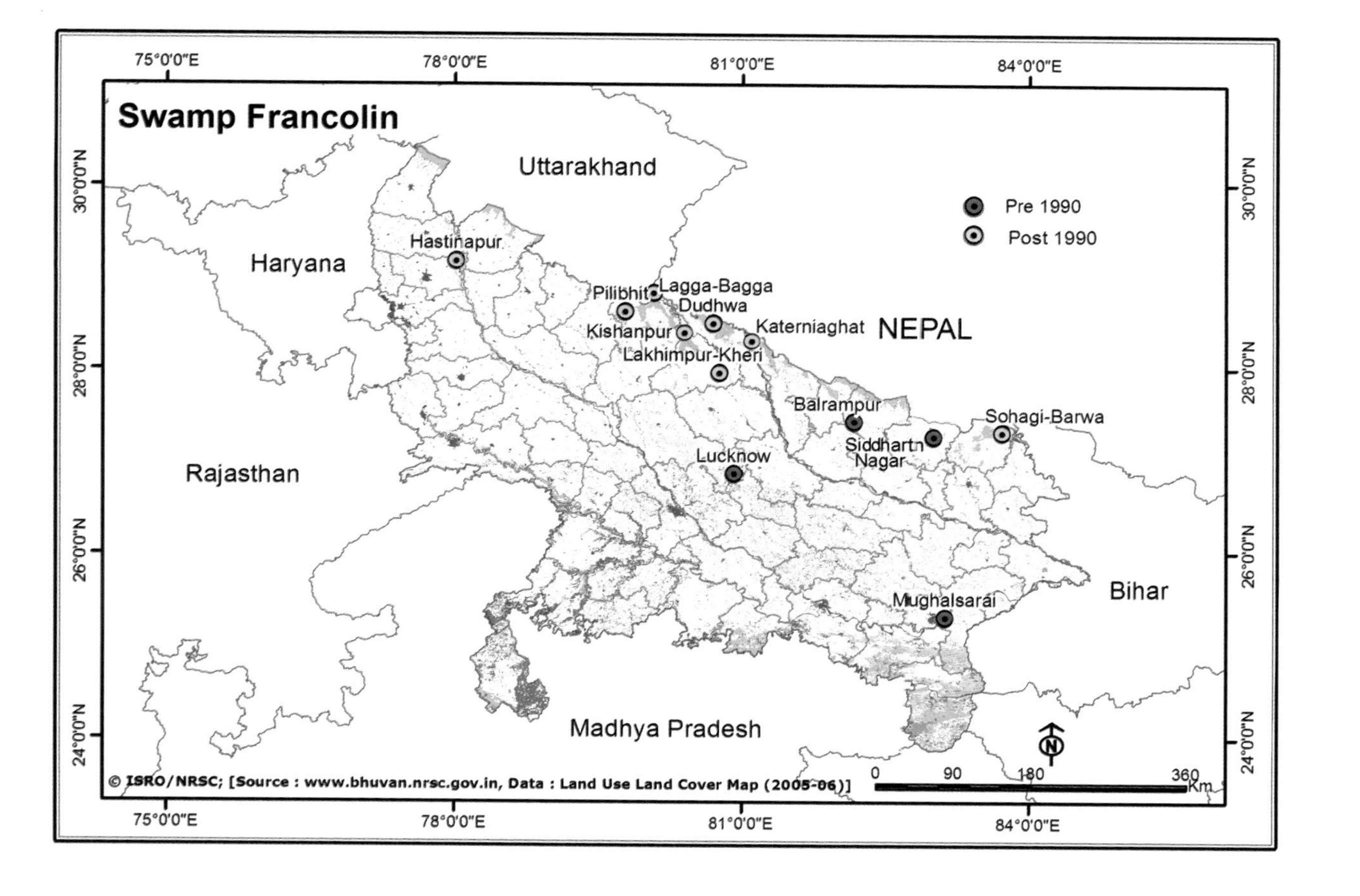

Swamp Francolin
Pre 1990
Post 1990
NEPAL
Uttarakhand
Haryana
Rajasthan
Madhya Pradesh
Bihar
Hastinapur
Pilibhit
Lagga-Bagga
Dudhwa
Kishanpur
Lakhimpur-Kheri
Katerniaghat
Balrampur
Siddhartn Nagar
Sohagi-Barwa
Lucknow
Mughalsarai
75°0'0"E
78°0'0"E
81°0'0"E
84°0'0"E
30°0'0"N
28°0'0"N
26°0'0"N
24°0'0"N
0
90
180
360
Km
© ISRO/NRSC; [Source : www.bhuvan.nrsc.gov.in, Data : Land Use Land Cover Map (2005-06)]

Distribution: The Swamp Francolin is endemic to the Ganga and Brahmaputra river basins, from the Terai in western Nepal through Uttar Pradesh, Bihar(?), West Bengal, Assam, and Arunachal Pradesh. Rahmani (2012) has given its historical and recent records in India. In this book, we give records from the Uttar Pradesh *terai* areas.

Records from Uttar Pradesh: Javed & Rahmani (1991) have reported it from 12 protected areas in northern India. In a survey funded by Birdquest/OBC in the Uttar Pradesh *terai*, Javed (2001) in March 1998 found it at 24 of 35 sites where dawn call-counts were conducted. Nine of the 24 sites are currently protected. The population estimate was 136 calling birds (each calling bird assumed to be a pair). **Pilibhit** and **Lakhimpur-Kheri** districts held the majority of calling birds, with a particularly large population of *c.* 50 calling birds in the grasslands bordering a man-made canal in Pilibhit Reserve Forest, and another major population in **Dudhwa** NP. It is seen in **Katerniaghat** and **Kishanpur** Wildlife Sanctuaries.

A series of small patches of habitat in the central and western Terai did not produce records of the species, perhaps due to their isolation from large source populations (Javed 2001). According to Harish Kumar Gularia (*pers. comm.* 2010), it is found in good numbers in the Chuka and Malsi blocks of **Mahauf** Range Forests and **Lagga-Bagga** block of **Barhi** Range Forests of **Pilibhit** RF. A pair was spotted in Hathikund area of Pipra range in **Suheldev** WLS, and a few times in Daebhar range of **Sohagi Barwa** WLS. It is presumed that the species was previously present in Hastinapur WLS, Meerut district (Bhargava 2012).

Ecology: As Uttar Pradesh is very important for the survival of Swamp Francolin, we are describing its ecology in detail based on the studies conducted during the last 20 years. The Swamp Francolin inhabits tall, wet, natural grasslands, particularly those dominated by *Phragmites*, *Arundo*, *Saccharum*, and *Narenga*, and also occurs (at lower densities) in wet agricultural areas which are dominated by sugar cane and paddy interspersed with natural vegetation (McGowan *et al.* 1995, Javed 1996, Javed *et al.* 1999, Iqubal *et al.* 2003, Baral 1998). It is predominantly known from the lowlands (generally <250 m), but moves to slightly higher altitudes during periods of high flood. Its range size is small compared with other Galliformes (Iqubal *et al.* 2003). Moderate levels of grazing and disturbance are not significantly detrimental to the species (Javed 2001).

Like most members of its family, the Swamp Francolin is a territorial bird, with pairs maintaining discrete territories. Call is generally described as *qua, qua, qua*, hence its popular name Kyah (Bengal) and *Koi Koi sorai*, or *Koira* (Assam). In north India, it nests in summer (March to June) just before the monsoon starts. Though monogamous, incubation is by the female alone. However, both parents help in chick rearing. Like other francolins, it feeds on grass seeds, small insects, termites, and shoots of crops. The species is generally shy and retreats to tall grasses on the slightest disturbance, but in protected areas it can be seen feeding on roads and clearances in the early morning and evening. It prefers to run and hide, but also flies over grassland to escape predators, settling down 20–30 m away. It is rather difficult to flush out a second time.

In a study of habitat use pattern, sighting of the Swamp Francolin did not show significant correlation with broad categories of tall, medium, and short grasslands, but showed significant preference for different grass associations (Javed *et al.* 1999). The maximum association was with *Scherostachya fusca* and *Saccharum* spp. Interestingly, significant correlation was observed between Swamp Francolin sightings and linear distance from a waterbody. The majority of sightings occurred within 200 m of the water source (Javed *et al.* 1999). Cattle grazing adversely affects the Swamp Francolin: it tolerates light grazing, but avoids medium to heavily grazed grassland patches. During burning of grasslands, the Swamp Francolin takes refuge in unburnt patches. Within a week after burning, it is found randomly distributed in burnt and unburnt patches.

Group size of the Swamp Francolin varies from 1–10, with the majority of sightings of adults being in pairs. Four is the maximum group size in adults. Bigger groups constitute a pair and chicks. On an average, five chicks per pair or per mother (range of group 2–8) were observed (Javed *et al.* 1999). Based on five nests studied, Iqubal *et al.* (2003) found that the mean clutch size was 5.4 (range 4–7).

Based on a study of 13 radio-tagged birds, Iqubal *et al.* (2003) found that, in and around Dudhwa NP, the home range size varied from 273 sq. m to 2,687 sq. m. Nesting also influences the home range. It was found that the smallest home range was of nesting birds and largest home range was of birds that were not nesting. At the home range level, birds appeared to favour tall sugar cane and grassland, whilst at the individual location level, grasslands and wet areas were most used.

Threats: Habitat destruction and deterioration are the two biggest threats to the Swamp Francolin. Its damp grassland habitat is under intense human pressure due to its productive soil which can be easily converted to rich agricultural fields (most have already been converted). Burning of grassland during the nesting period is another danger, along with overgrazing, cutting of grass, and conversion of natural grasslands to commercial plantation. Hunting for meat and trapping for sport (cock fighting) also occur, but possibly on a smaller scale now. The main demand is either for large (private) aviculture collections, including zoos, and local consumption as meat. Most people do not keep it for display due to fear of persecution. A small percentage is also kept by locals as call birds (decoys) and sometimes for cock fighting. Not less than 100 birds were recorded during market surveys in 1994 and 2006, a figure quite low as most meat trade goes undocumented (Ahmed 1997, 2002, and *in litt.* 2010).

According to Ahmed (2012), the collection points for organized trade are Lakhimpur-Kheri and Pilibhit, from where the birds reach Lucknow and Kanpur, from where they are further traded. A pair for aviculture collections or zoos can fetch up to Rs. 5,000. However, current surveys suggest that trapped birds may

SANJAY KUMAR

The Swamp Francolin breeds from March to June. Four to seven eggs are laid but generally two or three chicks are raised successfully

be sold for meat by local trappers at the grassroots level due to fear of being detected and persecuted during transportation to bigger cities.

Use of pesticides is unregulated in India and Nepal. Although no study has been conducted in these countries on the impact of pesticides on this species, agricultural pesticides may be affecting its numbers, either through direct mortality or reduction in potential food sources (invertebrates). Drying/flooding of its wet grassland habitat due to climate change and irregular fluctuations in water level could become a major threat in the future.

Conservation measures underway: The Swamp Francolin is covered by Schedule IV of the Indian Wildlife (Protection) Act, 1972. Although its specialised habitat (*terai* grassland) has largely disappeared, the surviving habitat is safe in the 14 IBAs and in PAs in India (Javed 2000, Javed *et al.* 1999, Islam & Rahmani 2004). Fortunately, it can survive in sugar cane fields adjoining natural grasslands, so it is not as threatened as earlier thought. Some restoration of *terai* grassland has taken place in North Pilibhit Reserve Forest due to seepage from canals. Another issue is increase in protection level and regulated grassland management practices in Pilibhit Reserve Forest which could not be suitable for this species (Harish Gularia *pers. comm.* 2010). The Terai Arc Project of WWF in India and Nepal has also recommended many measures for protecting the natural

The Swamp Francolin lives in a mosaic of tall and short grasses near water

ecosystem of the Terai. Good management practices in Katerniaghat WLS have resulted in the restoration of grasslands suitable for the Swamp Francolin, particularly in the Girijapur Barrage floodplains. Unfortunately, most of the suitable habitat in Hastinapur WLS has been destroyed during the last 10 years (Affif Khan *pers. comm.* 2010).

RECOMMENDATIONS

(1) Study the impact of the grass burning regime in protected areas such as Dudhwa, Kishanpur, and Katerniaghat, and take appropriate corrective measures.
(2) Map, inventorise, and protect all the grasslands suitable for this species outside protected areas, particularly those which are close to PAs/IBAs. Develop site-based management regimes that are compatible with the survival of the species in anthropogenic habitats and farming techniques.
(3) Generate a management plan for the whole range of the species to address the fragmentation of its habitat.
(4) Implement strict control on poaching and trapping.
(5) Explore the possibility of reintroduction in the restored grasslands of Narora, Uttar Pradesh.
(6) Initiate special enforcement and education drives in the known trade areas such as villages in Pilibhit, Lakhimpur-Kheri, and Sitapur districts.

Marbled Duck

Marmaronetta angustirostris (Mènètriés 1832)

ANAND ARYA

BirdLife International (2013) justifies keeping the Marbled Duck in the Vulnerable category as it has suffered a rapid population decline, evident in its core wintering range, as a result of widespread and extensive habitat destruction. It is a marginal species in Uttar Pradesh with very few records.

Field Characters: A small duck *c.* 48 cm, with an overall appearance of grey-brown flecked with creamy brown. It has a dark eye-patch and broad eye-stripe from eye to nape. A slight nuchal crest is present. It lacks speculum or wing mirror, unlike teals. Ventral portion is sullied white, more or less barred transversely with brown (Ali & Ripley 1987). Female slightly smaller. Juveniles are similar to adult, but with more off-white blotches. In flight, the wings look pale without a marked pattern, and lack speculum on the secondaries. Pale blotches on the mantle and wing coverts give a marbled effect to the plumage, hence the name.

Distribution: The Marbled Duck has a fragmented distribution in the Mediterranean, and western and southern Asia (BirdLife International 2001, 2013). In the Indian subcontinent, it is generally rare and local, with small numbers breeding and wintering in (chiefly southern) Pakistan and wintering in northwest India (with a few records from Assam and a report from Bangladesh) (Ali & Ripley 1987, Grimmett *et al.* 1999).

In **India**, it is a rare winter visitor to the northwest, north, and northeastern states. It has been reported from Jammu & Kashmir, Punjab, Haryana, Delhi,

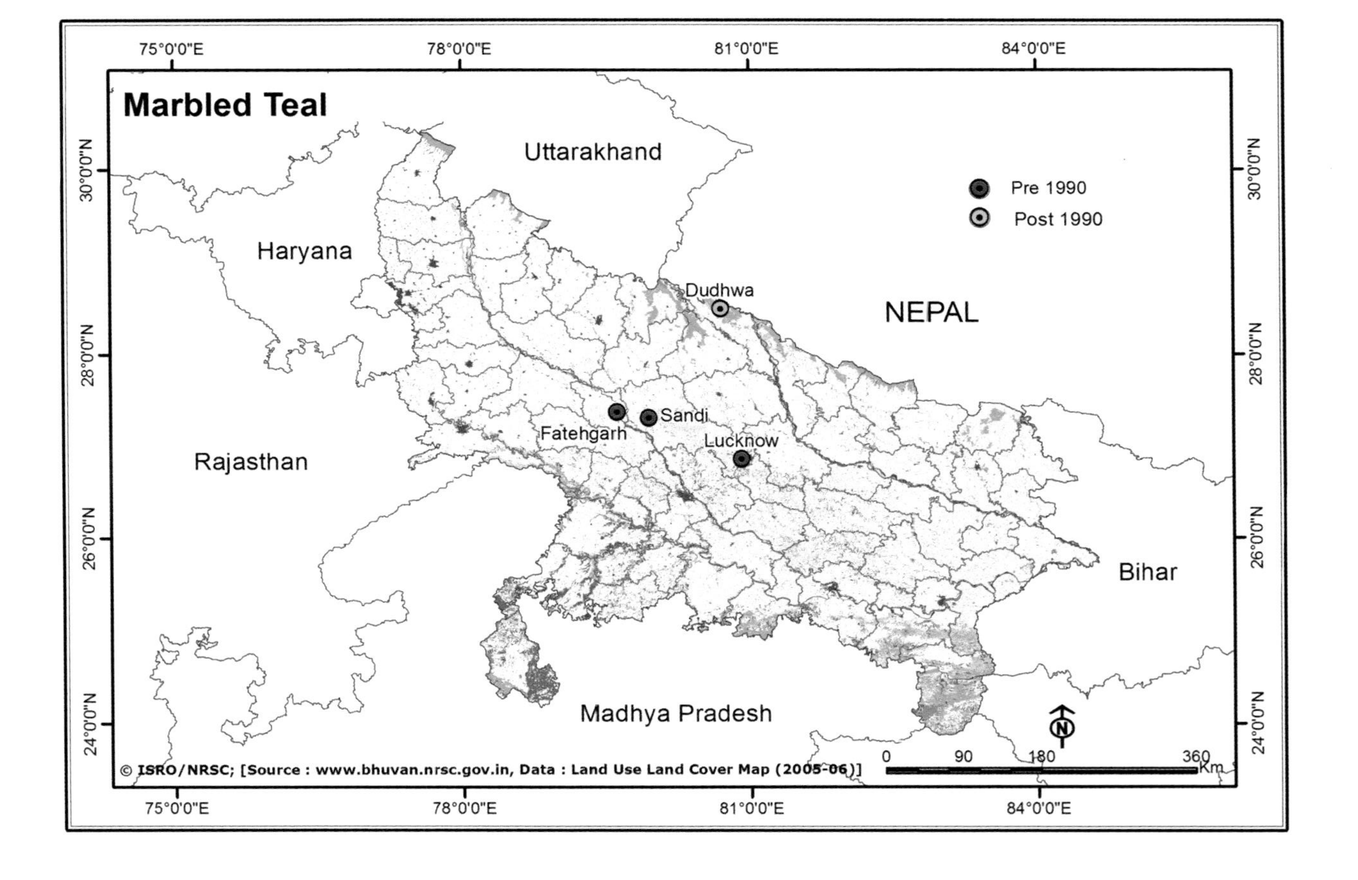
Marbled Teal
Pre 1990
Post 1990
Uttarakhand
Haryana
Rajasthan
NEPAL
Dudhwa
Sandi
Fatehgarh
Lucknow
Bihar
Madhya Pradesh
0 90 180 360 Km
75°0'0"E
78°0'0"E
81°0'0"E
84°0'0"E
30°0'0"N
28°0'0"N
26°0'0"N
24°0'0"N
© ISRO/NRSC; [Source : www.bhuvan.nrsc.gov.in, Data : Land Use Land Cover Map (2005-06)]

Gujarat, **Uttar Pradesh**, Bihar, Rajasthan, Maharashtra, West Bengal, and Assam (for details see Rahmani & Islam 2008).

Distribution in Uttar Pradesh: It is an extremely rare winter visitor to Uttar Pradesh, with less than 10 confirmed records during the last 150 years or more. For instance, Cunningham (1928) reported it to have been shot in **Roorkee** district between 1892–1905, while Anderson (1875) reported two from **Fatehgarh** district, and at least three in **Sandi** jheel in Hardoi district, of which one was shot. A specimen was captured and given to the **Lucknow** Museum by a wildfowler (Reid 1881). In recent years, Javed & Rahmani (1998) found it to be occasional in the wetlands of **Dudhwa** National Park, Lakhimpur-Kheri district. During the last few years, two pairs were sighted in January 2011, and again five of them in mid-February 2012, in the **Sandi** Bird Sanctuary, Hardoi district.

Ecology: Many studies have been conducted on this important threatened species and a Species Action Plan has been prepared. The results of these studies have been summarized by del Hoyo *et al.* (1992) and BirdLife International (2001, 2013). In **India**, the Marbled Duck is found on reedy and matted vegetation-covered jheels. As it is a rare migratory bird in Uttar Pradesh, we are not describing its ecology in detail.

Threats: It suffers from the usual biotic pressures faced all over the world by its wetland habitats: drainage for agricultural or industrial or hydrological purposes, pollution, and excessive livestock grazing. As it breeds in tall dense reeds, reed-cutting, reed-burning, and grazing are a great threat during the nesting season. It is also poached in most of its range, particularly in the Mediterranean and Middle Eastern countries where hunting tradition is still very strong. In India, it faces all these problems but probably to a lesser extent.

Conservation measures underway: It is listed in Schedule IV of the Indian Wildlife (Protection) Act, 1972, and also in CMS Appendix I and II. In India, in recent years, it has been reported from some IBAs/PAs (Islam & Rahmani 2004). Hunting and trapping are still a threat in some parts of India.

RECOMMENDATIONS

As it is an irregular winter migrant in India, we cannot do much to protect this species significantly. BirdLife International (2013) has given recommendations for its protection in all range states. Here we give some general recommendations for India and the state:

(1) Conduct regular surveys and monitoring in wetlands in Uttar Pradesh.
(2) Conduct research on its wintering ecology.
(3) Protect habitats at all sites regularly holding the species.
(4) Strictly implement ban on poaching and trapping.
(5) Increase public awareness regarding wetland protection.

Sarus Crane

Grus antigone (Linnaeus 1758)

SANJAY KUMAR

Sarus is the State Bird of Uttar Pradesh and probably 60% of its population in India is found in this state. BirdLife International (2013) and IUCN consider Sarus Crane as Vulnerable as it has suffered a rapid population decline, which is projected to continue, as a result of widespread reductions in the extent and quality of its wetland habitats, poisoning, and pollutants.

Field Characters: Sarus Crane is the tallest flying bird in the world. It stands 152–156 cm tall. Adults are grey overall, with whiter mid-neck and tertials, mostly naked red head and upper neck, blackish primaries, but mostly grey secondaries, and reddish legs that are bright during the breeding season and pale outside the breeding season. Females are supposed to be slightly smaller, but sometimes this difference is imperceptible. *Grus antigone sharpi* is a more uniform, darker grey. Juvenile has feathered buffish head and upper neck, and duller plumage with brownish feather fringes. The bare red skin of the adult head and neck is brighter during the breeding season. This skin is rough and covered by papillae, and a narrow area around and behind the head is covered by black bristly feathers.

Distribution: The Sarus Crane has three disjunct populations — in the Indian subcontinent, Southeast Asia, and northern Australia. Meine & Archibald (1996) estimated a global population of 25,000–37,000 individuals, which was brought down to 15,000–20,000 individuals by Archibald *et al.* (2003). The nominate

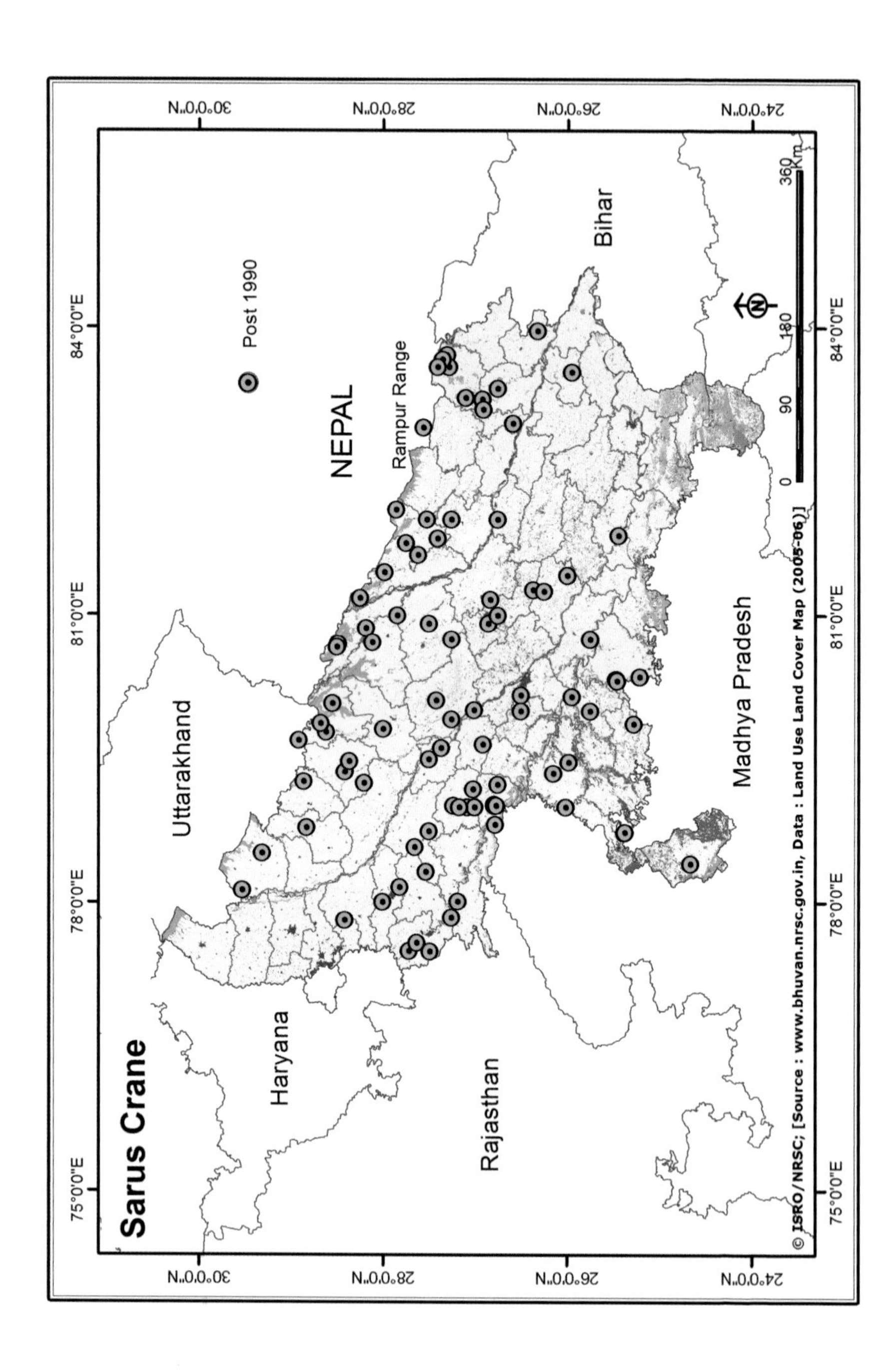
Sarus Crane
Post 1990
NEPAL
Rampur Range
Uttarakhand
Haryana
Rajasthan
Madhya Pradesh
Bihar
30°0'0"N
28°0'0"N
26°0'0"N
24°0'0"N
75°0'0"E
78°0'0"E
81°0'0"E
84°0'0"E
0 90 180 360 Km
© ISRO/NRSC; [Source : www.bhuvan.nrsc.gov.in, Data : Land Use Land Cover Map (2005-06)]

Both parents help in incubation and chick rearing. Generally two eggs are laid, very rarely three

subspecies *Grus antigone antigone* (*c.* 8,000–10,000 birds) inhabits northern and central India, Nepal and Pakistan (although now thought to be extinct as a breeding species there: Archibald *et al.* 2003), with occasional vagrants in Bangladesh. Its range has contracted towards the north and west of the Subcontinent and its population is considered to be in decline (Sundar *et al.* 2000, Sundar & Choudhury 2003, Archibald *et al.* 2003). Based on his surveys and intensive studies, Sundar (2005a) estimates 2,500–3,000 Sarus Crane in Etawah and Mainpuri districts of Uttar Pradesh. This equals *c.* 30% of the estimated global population of *G. a. antigone* (8,000–10,000 individuals). However, according to the census of the Forest Department, its population is between 15,000 and 18,000 in the state.

Sundar *et al.* (2000) conducted a comprehensive district-level survey to determine the distribution, demography, and status of the Sarus Crane, between June 1998 and March 1999 in the states of Jammu & Kashmir, Himachal Pradesh, Punjab, Haryana, Rajasthan, Gujarat, Uttar Pradesh, Madhya Pradesh, Bihar, West Bengal, and Maharashtra. Recently, Rahmani (2012) has given the latest Sarus numbers from all over India. In this book, we described its distribution in Uttar Pradesh.

Records from Uttar Pradesh: Sarus is fairly common throughout the state and mainly concentrated in the central and northwest parts which can be termed as its core territory within the Gangetic plains. Various low-lying areas interspersed with agricultural fields in the adjoining areas provide a matrix which boasts some of the finest habitats for this bird. We have two distinct areas or clusters. One is formed by Etawah, Mainpuri, Etah, and Aligarh districts, while the other is comprised of Lakhimpur-Kheri, Shahjahanpur, Sitapur, Hardoi, and Barabanki districts, where large congregations can be seen throughout the year except in the monsoon. In Sitapur district, out of 30 wetlands taken up for monitoring and restoration, 22 have shown the presence of Sarus. A seasonal survey revealed the presence of 323 Sarus (including 31 juveniles) (Kumar & Srivastav 2011). In Barabanki, alongside a canal on Dewa road, more than 200 Sarus can be found within an area of 10 sq. km. At most of the sites, there is no immediate threat as Sarus is considered sacred by the locals. But alteration of land use is a threat, as we noticed in Raghunath-ka-purva, a hamlet near Lucknow, where a shallow waterbody was drained for farming, resulting in the disappearance of a breeding pair (Amit Mishra *pers. comm*. 2013). Sporadic egg stealing and predation by dogs and crows were reported at some places. Some of the protected wetlands such as Nawabganj, Sandi, Samaspur, Patna, and Lakh-Bahosi have regular and permanent breeding pairs. Sarus is also reported from Baruasagar (Jhansi), Banda, Kanpur, and Fatehpur districts (Amit Mishra *pers. comm*. 2013); Hakimpur wetlands of Narora Atomic Power Plant-Exclusion zone (Raja Mandal *pers. comm*. 2013).

Sarus is commonly seen in agriculture fields, both in paddy and fallow land on both sides of railway tracks. It is almost certain to obtain sightings of this bird while travelling by train from Lucknow to Moradabad or from Lucknow to Agra via Kanpur and Etawah.

Lakhimpur-Kheri district is famous for Sarus. It is not uncommon to see groups of 40–50 Sarus, particularly after the monsoon. In January 2010, 142 Sarus were seen in a dry wetland near Shohratgarh *tehsil* of Siddharthnagar district (Fazlur Rahman & Mohammed Amir *pers. comm.* 2013). Another important sighting was of 177 birds in February 2010 in a crop field near Ucholiya Forest Rest House in Shahjahanpur district (Fazlur Rahman & Sanjay Kumar *pers. comm*. 2013).

Sarus has been reported from the following PAs/IBAs in Uttar Pradesh (Rahmani 2012): Bakhira WLS (Sant Kabir Nagar), Dudhwa National Park (Lakhimpur-Kheri [Palia *tehsil*]), Hastinapur WLS (Muzaffarnagar, Meerut, Ghaziabad, Bijnor, Jyotiba Phule Nagar), Katerniaghat Sanctuary (Bahraich), Kishanpur Wildlife Sanctuary (Lakhimpur-Kheri), Kudaiyya Marshland (Mainpuri), Kurra Jheel (Etawah, Mainpuri), Lagga-Bagga Reserve Forest (Pilibhit), Lakh-Bahosi Bird Sanctuary (Farrukhabad), Narora (Bulandshahr, Badaun), National Chambal Wildlife Sanctuary (Agra, Etawah), Nawabganj Bird Sanctuary (Unnao), Parvati Aranga WLS (Gonda), Patna Bird Sanctuary (Etah), Payagpur [Bagheltal] Jheel (Bahraich), Saman WLS (Mainpuri), Samaspur Bird Sanctuary (Raebareli), Sandi (Hardoi), Sauj Lake (Mainpuri), Sarsai Nawar Lake (Etawah), Sheikha Jheel

Nearly 60% of India's Sarus population is found in Uttar Pradesh. Other states with good Sarus numbers are Rajasthan, Gujarat, and Madhya Pradesh

(Aligarh), Sitadwar (Bahraich), Sohagi Barwa WLS (Gorakhpur), Sur Sarovar Bird Sanctuary (Agra), and Surha Taal WLS (Ballia)

Sarus Crane can easily be spotted in some lesser known wetlands and agricultural fields in the districts of Etawah, Mainpuri, Etah, Aligarh, Hathras, Auraiya, Kannauj, Kanpur Dehat, Lakhimpur-Kheri, Bahraich, Gonda, Shahjahanpur, Sitapur, Barabanki, and Hardoi. The last Sarus census conducted by the Forest Department in Ocober 2012 revealed active Sarus Crane breeding in good numbers in the districts of Moradabad, Bijnor, Rampur, Amroha, and Badaun in smaller wetlands between 2–5 hectares. In Moradabad alone, a single-day census on October 2, 2012 revealed a total of 86 cranes including 22 chicks/juveniles. During the winter season between December and March, Sarus congregates in large numbers in some lesser known places such as Machecha in Mohammadi *tehsil* of Lakhimpur-Kheri (>150), Campierganj in Gorakhpur (>220), wetlands in Hargaon (Sitapur), Oyle in Lakhimpur-Kheri (>100), and a few smaller wetlands in Shahjahanpur and Hardoi (>100).

Nikhil Shinde and Rajat Bhargava of BNHS during their bird surveys of the Terai from November 2013 to April 2014 have seen Sarus in the following areas (number and date given in bracket): Poorvi Sohelwa Range, Soheldev WLS (2 on December 28, 2013); near Nanpara, Dist. Bahraich (2 on December 12, 2013); South Chowk Range, Sohagi Barwa WLS (2 on February 7, 2014); Madhaulia Range, Sohagi Barwa WLS (2 on February 6, 2014); Belhiya Bazaar, Near North Chowk,

Sarus is traditionally protected and venerated by villagers in India

Sohagi Barwa WLS (35 on February 7, 2014); Mahouf Range, Pilibhit Reserve Forest (7 on April 4, 2014); Musva Taal, Dist. Bahraich (2 on February 11, 2014); Narayanpur village, *tehsil* Ikauna, Dist. Shravasti (3 on January 17, 2014); and Rafaddur *thana*, Dist. Lakhimpur (5 on February 3, 2014).

Ecology: The ecology and behaviour of Sarus have been extensively studied by Sundar *et al.* (2000), Sundar & Choudhury (2003, 2005), Sundar (2009), Kaur (2008), and Kaur *et al.* (2008).

The Sarus uses open wet and dry grasslands, agricultural fields, marshes, and jheels for foraging, roosting, and nesting. Wetlands, even very small ones close to roads and human habitation, are used to construct nests (Sundar 2009). For foraging, Sarus usually uses crop fields to a lesser extent and prefers feeding in wetlands. It is omnivorous, feeding on a variety of roots and tubers as well as invertebrates and amphibians. In drought years, it concentrates in the remaining wetlands, particularly in summer.

Sarus has a long breeding season, starting with the onset of monsoon (July) and extending to October-November. In some suitable places, it also breeds in February-March, e.g., seepage wetlands of Kota canal (Kaur 2008, Kaur *et al.* 2008). Both parents select the nest site and help in nest building. Clutch size is usually one or two eggs, but mostly one chick is successfully raised. The juvenile moves with the parents for almost a year, till the next breeding season. In areas with perennial water supply through wetlands and irrigation canals, pairs maintain discrete territories throughout the year. The Sarus makes very large nests (up to 2 m in diameter). Most nests are constructed entirely of grasses and other wetland plants.

Studies on the breeding biology of Sarus in India have shown that very few have a clutch size of three, and breeding pairs characteristically raise one or two chicks each year (Sundar & Choudhury 2003). However, Sundar (2006) found a pair on December 12, 1998 in its breeding territory foraging with three immatures near Bhartana town in Etawah district, and another pair with three newly fledged chicks in Kheda district of Gujarat on January 26, 2006. Kaur *et al.* (2008) recorded for the first time three juveniles during the dry season (February-May) nesting in Ayana wetland of Baran district in Rajasthan.

Although sexes are similar in Sarus, the adult male is slightly larger than the female (clearly seen when both are together). When they call in unison, the male droops its primaries and touches the secondaries over its back, which can be used for rapid sexing (Sundar & Choudhury 2005).

Threats: In India, Sarus is considered a sacred bird, so hunting is not the main problem: it is habitat destruction and habitat alteration which are taking their toll. Wetlands are under tremendous pressure from human use, drainage and conversion to agriculture, housing colonies, and even construction of highways. A recent hazard has emerged in the form of poisoning (Muralidharan 1992, Kaur & Nair 2008). Collision with power lines may be a significant threat in parts of its range, with observations from India suggesting that 2.5–20% of some populations are affected by such stochastic occurrences (Sundar *et al.* 2000, Sundar & Choudhury 2001, Rana & Prakash 2004). In some areas the annual mortality of mostly non-breeding birds is almost 1% of area's total population (Sundar & Choudhury 2006). The other major reason for the decline in Sarus numbers is egg mortality (Mukherjee *et al.* 2001, Kaur & Choudhury 2003, Sundar 2009). Predation on eggs is largely by crows (Ramachandran & Vijayan 1994, Sundar 2009). An instance of a young chick predated by Marsh Harrier was recorded for the first time from Kota in Rajasthan (Kaur & Choudhury 2005).

According to Sundar (*pers. comm.* 2010), chick predation by dogs and egg predation by corvids is increasing, as their populations increase following the decline of vultures on the Indian subcontinent.

According to intensive studies on bird trade conducted by TRAFFIC India/WWF-India since 1992, it is alarming to note that in certain villages around Lucknow, Kanpur, Allahabad, Varanasi, and Rampur, Sarus chicks are highly valued and systematically collected and traded by the Baheliya, Pathmani, and Mirshikar tribes for pets, zoos, and aviculture throughout India, despite the blanket ban on wild bird trading since 1990–91 (Ahmed 1997, 2002, 2012; Ahmed & Rahmani 1996). Infrequent raids conducted by the UP Forest Department resulting in seizure of up to 10 Sarus Crane have proved this finding from time to time. Further, in certain areas in Kannauj, Unnao, and Bhargarmo, adult Sarus are also caught with plastic leg nooses placed near the nest and the captured birds are traded for meat which can fetch up to Rs 1,500–2,000 per bird in weekly village markets.

Conservation measures underway: The Sarus Crane is listed in Schedule IV of the Wildlife (Protection) Act, 1972 and in CITES Appendix II and CMS Appendix II. It is the State Bird of Uttar Pradesh. The Uttar Pradesh Government has established the Sarus Protection Society. Sarus is present in many PAs/IBAs, but the majority of the population is found in cultivated lands, protected largely by farmers who are tolerant of these sacred birds.

RECOMMENDATIONS

(1) Annual state-wise surveys in summer and winter, involving a large number of volunteers and forest officials should be conducted under the supervision of Sarus experts so a scientific database is developed.

(2) Detailed research on local and seasonal movements of Sarus Crane in both wet and dry season nesting period is required by banding more juveniles and satellite tracking (PTTs).

(3) Hydrological data on individual wetlands that harbour Sarus Crane, and that of large habitats such as river basins and sub-basins are urgently required, particularly in view of the changing rainfall regimes that appear to be a consequence of global climate change.

(4) Removal of encroachments from wetlands and restoration of the marshes along wetlands are required by declaring them as Conservation Reserves or Community Reserves.

(5) Sarus is mostly present in agricultural fields and hence people's protection is of utmost significance. Destruction of Sarus nests, stealing of eggs, and occasional hunting are direct threats. Involving the local communities around the breeding sites through education and awareness campaigns on a prolonged basis is essential.

(6) More research on the cranes' role in agriculture, e.g., in pest control, needs to be carried out.

(7) Interdepartmental discussions and policy regulation for the protection of village ponds and wetlands are also necessary.

Pallas's Fish-eagle

Haliaeetus leucoryphus (Pallas 1771)

RISHAD NAOROJI

IUCN and BirdLife International (2013) list Pallas's Fish-eagle in the Vulnerable category as this species has a small, declining population as a result of widespread habitat loss, degradation, and disturbance of wetlands and breeding sites throughout its wide range.

Field Characters: A large (76–84 cm), dark brown eagle with pale golden brown head and neck, and blackish tail with a broad, white central band. The band is particularly conspicuous in flight. Juvenile more uniformly dark, with all-dark tail, but in flight shows strongly patterned underwing, with whitish band across coverts and prominent whitish primary flashes. Female is larger. Voice is a loud, guttural *kha-kha-kha-kha* or *gao-gao-gao-gao*, and sometimes high-pitched, excited yelping. Hoarse, guttural, continuous *kook-kook-kook* is the commonest call.

Distribution: Pallas's Fish-eagle has a wide range from Central Asia, southern Russia, east Mongolia and China, northern India, Pakistan, Bhutan, Bangladesh, to Myanmar. In India, we have a small resident population which is augmented in winter by migrant birds from temperate regions. The main breeding populations

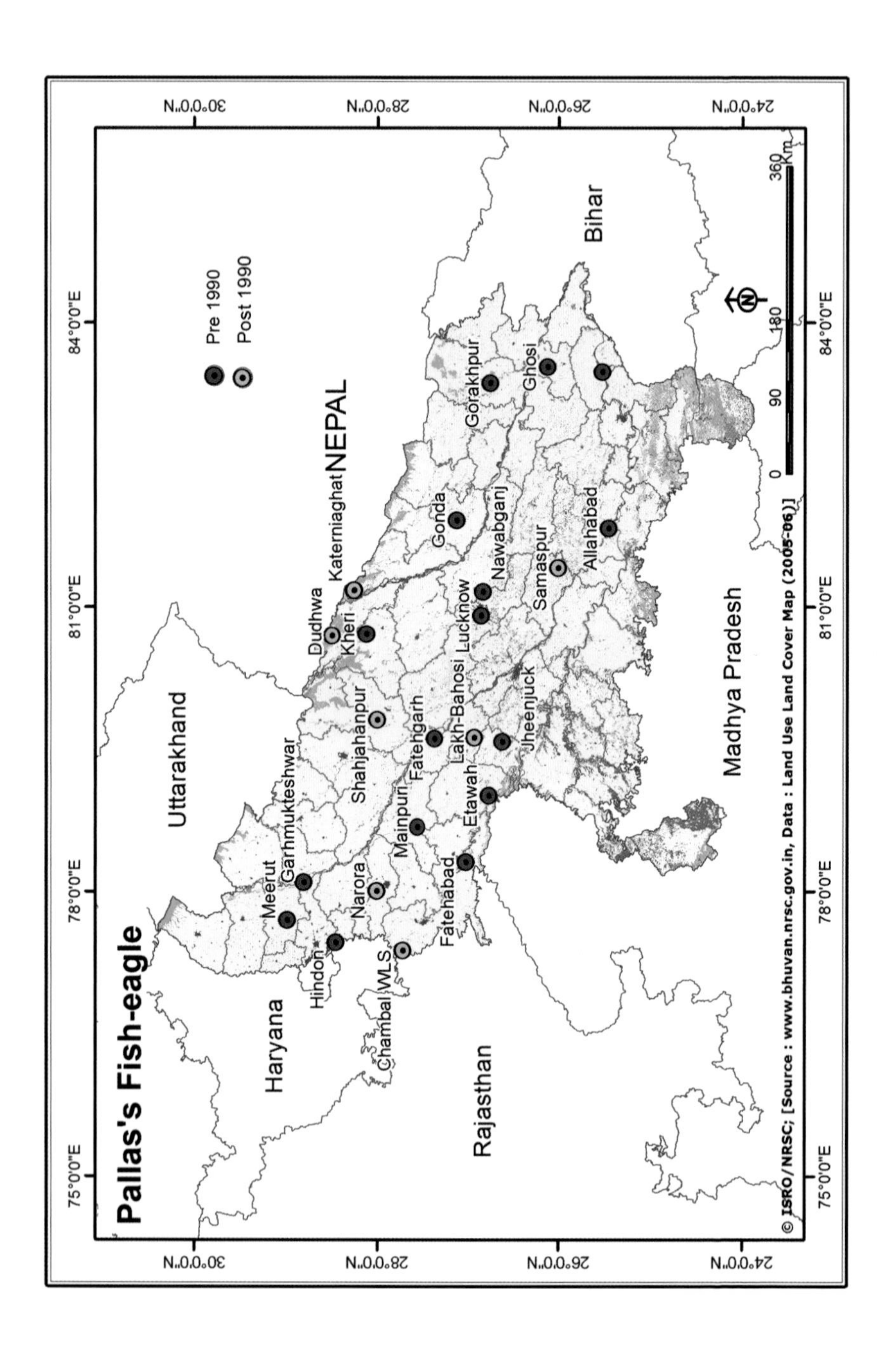

Pallas's Fish-eagle
Pre 1990
Post 1990
NEPAL
Uttarakhand
Haryana
Rajasthan
Madhya Pradesh
Bihar
Meerut
Garhmukteshwar
Hindon
Narora
Chambal WLS
Fatehabad
Mainpuri
Etawah
Fatehgarh
Lakh-Bahosi
Shahjahanpur
Jheenjuck
Kheri
Dudhwa
Katerniaghat
Lucknow
Nawabganj
Samaspur
Allahabad
Gonda
Gorakhpur
Ghosi
0 90 180 360 Km
30°0'0"N
28°0'0"N
26°0'0"N
24°0'0"N
75°0'0"E
78°0'0"E
81°0'0"E
84°0'0"E
© ISRO/NRSC; [Source : www.bhuvan.nrsc.gov.in, Data : Land Use Land Cover Map (2005-06)]

Most of the wetlands of Uttar Pradesh suffer from unregulated and intensive fishing which does not leave enough fish for Pallas's Fish-eagle and other piscivorous species

are believed to be in China, Mongolia, and the Indian subcontinent. The population is likely to be <10,000 mature individuals (BirdLife International 2013).

In India, it is still reported from larger rivers and wetlands in the north and northeast [see Naoroji (2007) for recent records]. Rahmani (2012) has given important records from India. Here we give records from the state.

Records from Uttar Pradesh: It has been reported from the following PAs/ IBAs: **Dudhwa** National Park, **Katerniaghat** Wildlife Sanctuary, **Lakh-Bahosi** Bird Sanctuary, **National Chambal** WLS, and **Samaspur** Bird Sanctuary. A nest was observed in use in **Katerniaghat** WLS for the last 20 years, but now it is not seen. It was also previously reported from **Hastinapur** WLS (Rai 1983) but needs recent confirmation (Bhargava 2012).

Two more confirmed records are available as two birds were reported from Hakimpura wetlands, **Narora** in February 2008 (P.D. Mishra & Raja Mandal *pers. comm.* 2013) and a pair was reported on a bare tree growing near a forest stream in **Shahjahanpur** range of Shivalik Forest Division. They could be nesting in both these places. During bird status survey of **Samaspur** WLS from July 2008 to March 2009, a pair of Pallas's Fish-eagle was found breeding (Rahmani 1992). Most of these records are two decades old. Now it is not present in many areas.

Ecology: Pallas's Fish-eagle is invariably found near water, mainly large wetlands, rivers, and jheels, from lowlands to 5,000 msl. It feeds mainly on fish, which are sometimes heavier than it can lift, and small waterfowl. It is well known for pirating prey from Osprey, Marsh Harrier, and Brahminy Kite, continuously harassing them till the prey is stolen. It is even reported to rob fish from Otters *Lutra lutra*, frightening them off to nab their prey (Lahkar 2000).

In heronries, it feeds on chicks of cormorants, Oriental Darter, egrets, ibises, and Asian Openbill, and is capable of killing birds as large as the Common Crane and Bar-headed Goose. In earlier literature it was mentioned that Pallas's Fish-eagle takes a heavy toll of young Bar-headed Geese in the lakes of Ladakh, but during five visits to these lakes from 2005 to 2008, no Pallas's Fish-eagle was seen. It generally nests on large trees near water. In India, the nesting season is from October to February. For details of its ecology and behaviour, see Naoroji (2007).

Birds breeding in northern climes are migratory and leave the area by October, and first breeders return by end-March. In north India, both resident and migratory birds are seen in winter, when the resident birds breed while adults and immatures arrive from Mongolia, China, and Central Asia. After breeding, the resident birds move from the hot lowlands to cooler higher areas.

Threats: Pallas's Fish-eagle suffers the same problems as almost all waterbirds (and being at the apex of the food pyramid, it is the first to disappear): habitat degradation through pollution, spread of Water Hyacinth *Eichhornia crassipes*, overfishing, felling of large trees near wetlands, and agriculture pesticides and industrial runoff to wetlands.

Conservation measures underway: It is listed in Schedule IV of the Indian Wildlife (Protection) Act, 1972, and also CITES Appendix II and CMS Appendix II. It occurs in many PAs/IBAs in India (Islam & Rahmani 2004).

RECOMMENDATIONS

(1) Conduct surveys in the whole state to establish its status, distribution, and threats.
(2) Study its movement and dispersal through ringing/tagging and satellite telemetry.
(3) Pesticide level in prey species should be monitored, and if high, remedial measures taken.
(4) Control Water Hyacinth at important breeding/feeding sites.
(5) Protect remaining nest trees and re-establish them around wetlands. Protect nesting sites (and adjacent feeding sites) from disturbance.
(6) Organic farming should be encouraged around important wetland nesting sites.

Greater Spotted Eagle

Aquila clanga Pallas 1811

NIRANJAN SANT

IUCN and BirdLife International (2013) list Greater Spotted Eagle as Vulnerable, as it has a small population which appears to be declining owing to extensive habitat loss and persecution.

Field Characters: A very dark eagle, 62–74 cm, invariably found near large jheels and wetlands. On perch at close range, the best way to separate it from other Aquilas is by its round nostrils and the gape line stopping at the centre of the eye. This facial characteristic shows specific variation in all Aquila eagles which are sometimes found in the same area in winter. On closer examination, the adult is dark brown (not black), with purplish or maroon sheen on mantle, and slightly paler ventral side. It also has slightly pale flight feathers, with underwing coverts generally darker than flight feathers. Sexes alike, but female is larger. Juveniles paler, with back and wings sparsely spotted or streaked with buff or white. For age-related plumage differences, see Naoroji (2007). While gliding, it often depresses its "hands". It can be confused with other Aquila eagles, but its habit of living near water helps to identify it.

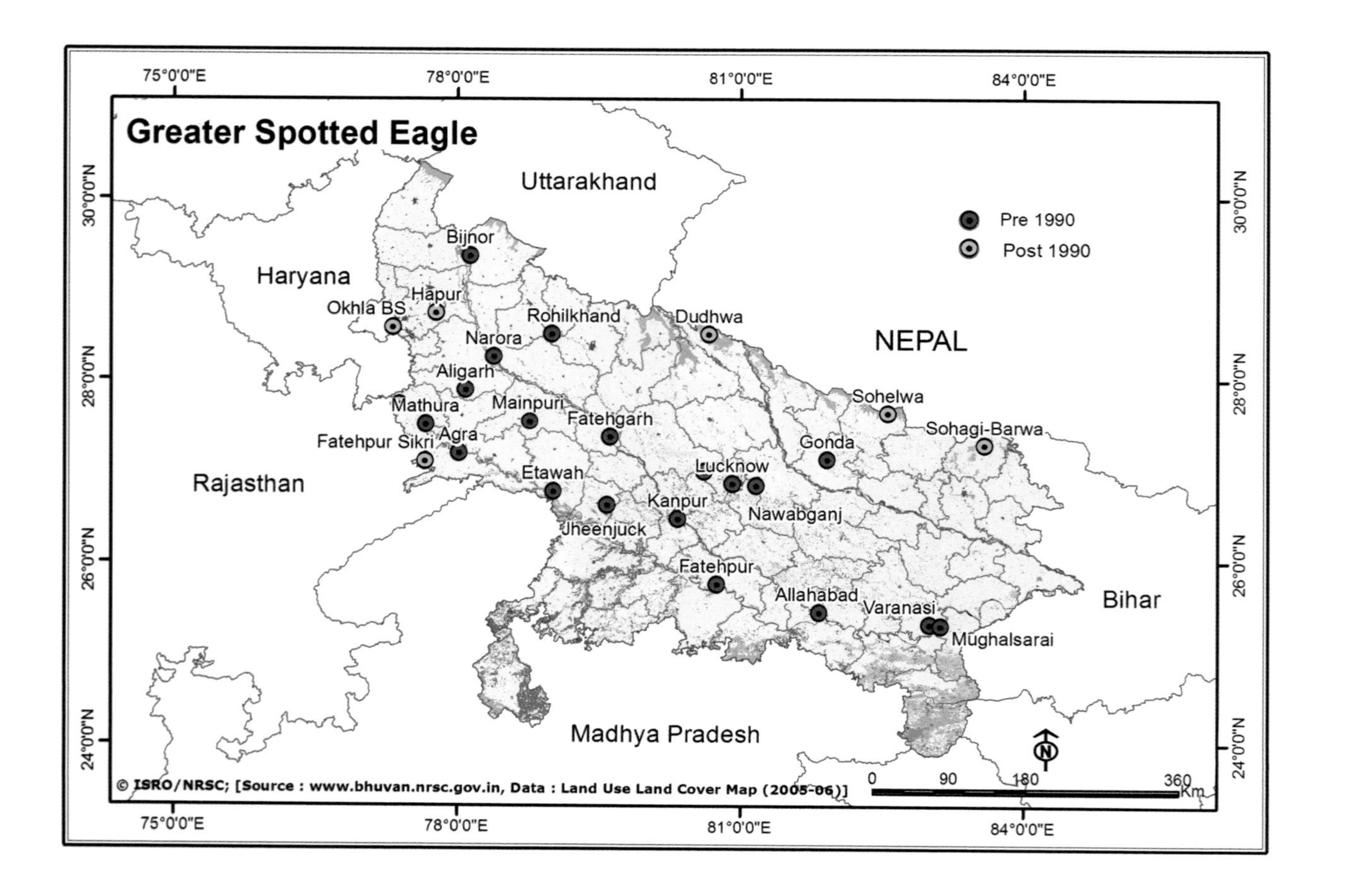

Greater Spotted Eagle
Pre 1990
Post 1990
Uttarakhand
Haryana
NEPAL
Rajasthan
Madhya Pradesh
Bihar
Bijnor
Hapur
Okhla BS
Rohilkhand
Dudhwa
Narora
Aligarh
Mathura
Mainpuri
Fatehgarh
Fatehpur Sikri
Agra
Sohelwa
Sohagi-Barwa
Gonda
Lucknow
Etawah
Kanpur
Nawabganj
Jheenjuck
Fatehpur
Allahabad
Varanasi
Mughalsarai
75°0'0"E
78°0'0"E
81°0'0"E
84°0'0"E
30°0'0"N
28°0'0"N
26°0'0"N
24°0'0"N
0 90 180 360 Km
© ISRO/NRSC; [Source : www.bhuvan.nrsc.gov.in, Data : Land Use Land Cover Map (2005-06)]

Distribution: In India, the Greater Spotted Eagle is a regular but uncommon winter visitor. It breeds in eastern Europe, Russia, and Central Asia, Mongolia and China. Passage migrants or wintering birds are seen in many countries (del Hoyo *et al.* 1994, Ferguson-Lees & Christie 2001). Its wide range appears to be deceptive, as it has fragmented populations which are undergoing an overall decline.

Rahmani (2012) has compiled major recent records in India. Here we give specific records for UP.

Records from Uttar Pradesh: Manoj Sharma (*pers. comm.* 2013) has recorded this species in winter in many places such **Garhmukteshwar** in Hapur district, **Okhla** Bird Sanctuary in Gautam Buddh Nagar district, **Fatehpur Sikri** in Agra district, and at River **Chambal**. Javed & Rahmani (1998) have recorded it as occasional at **Dudhwa** NP.

Nikhil Shinde and Rajat Bhargava of BNHS, during their surveys of birds in Sohagi Barwa and Soheldev WLSs, have seen one bird on December 22, 2013 in the Rampur Range, **Soheldev** and one bird on February 2, 2014 in South Chowk Range, **Sohagi Barwa**. It is likely to be present in more areas, particularly in wetlands but overlooked.

Ecology: The Greater Spotted Eagle is invariably found near water where it sits and waits for hours for its prey. These are waterfowl, particularly sick and injured, and chicks from heronries. In some parts of the world it mainly feeds on terrestrial prey — in wet grasslands, it feeds on amphibians and small mammals. It is also found on rubbish dumps and in mangroves. Its diet is highly variable. Information on its ecology in India has been compiled by Naoroji (2007).

Threats: In India, drainage and degradation of wetlands is the biggest threat to this species, and all waterbirds in general.

Conservation measures underway: Like all large birds of prey, the Greater Spotted Eagle is protected under the Indian Wildlife (Protection) Act, 1972 and listed in Schedule I. It is listed in CITES Appendix II, and CMS Appendix I and II. It is found in many IBAs/PAs and other sites in India (Islam & Rahmani 2004).

RECOMMENDATIONS

(1) The most urgent need is for regular surveys to assess its status, range, and population trends in the state.

(2) Identify the wetlands and wet grasslands where this bird winters in UP.

(3) Maintain, improve, and protect large wetlands through appropriate legislations and community conservation.

(4) Develop simple identification literature.

Indian Spotted Eagle

Aquila hastata (Lesson 1831)

VEER VAIBHAV MISHRA

BirdLife International (2013) justifies the inclusion of *Aquila hastata* in the Vulnerable category as it has a small and declining population. However, this statement has to be examined based on recent observations from Karnataka and Kerala. The species is regularly seen in small numbers over the whole Deccan plateau and a few wetlands in Kerala. This shows that it has a much larger range and perhaps a larger population than hitherto known.

Field Characters: A medium-sized, slim eagle with small bill and long gape extending to the middle of the eye, and round nostrils. It is supposed to have the widest gape among all Aquilas. Adult birds are drab brown and unspotted, with yellowish brown eyes and often paler brown vent and tarsi. The name Spotted is quite misleading as adult birds do not have spots. It has short, broad wings (wingspan 150 cm). In proportion to the Greater Spotted Eagle *Aquila clanga*, the Indian Spotted Eagle has a longer tail. The legs appear longer and thinner due to the tarsi being less thickly feathered, because of which it was also known as Long-legged Eagle. The head is large in relation to body size. First year juveniles have brown eyes and pale flecks on nape (no rufous patch), upper back, and wing-coverts; larger pale spots on median coverts and narrow pale

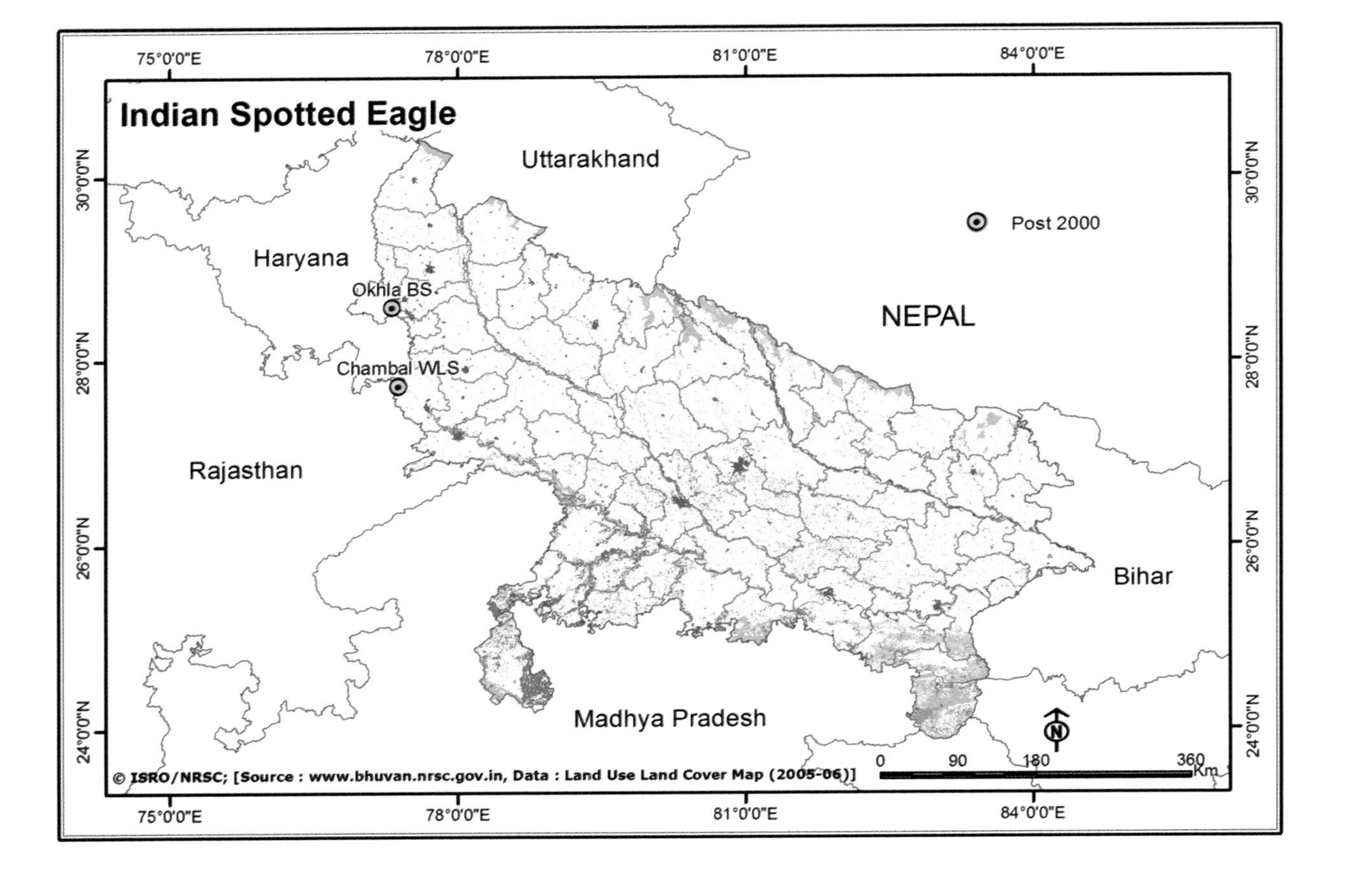
Indian Spotted Eagle
Uttarakhand
Haryana
Okhla BS
Chambal WLS
Rajasthan
NEPAL
Post 2000
Bihar
Madhya Pradesh
75°0'0"E
78°0'0"E
81°0'0"E
84°0'0"E
30°0'0"N
28°0'0"N
26°0'0"N
24°0'0"N
0
90
180
360
Km
© ISRO/NRSC; [Source : www.bhuvan.nrsc.gov.in, Data : Land Use Land Cover Map (2005-06)]

tips to tertials, and are often heavily streaked below (Rasmussen & Anderton 2005). For more details of age-related plumage variation, see Naoroji (2007).

Distribution: The Indian Spotted Eagle is a semi-endemic species, chiefly found in India, but also recorded in Pakistan, Nepal, Bangladesh, and Myanmar. According to BirdLife International (2013), sightings of Aquila in Cambodia almost certainly refer to this species, indicating that a population may persist in the dry deciduous dipterocarp forest mosaic in parts of Indochina.

It is present in drier biotope from north India to central and western regions, and there is probably a separate population in Karnataka, Tamil Nadu, and part of Andhra. For detailed distribution in India, see Naoroji (2007). It is reported from many PAs/IBAs such as Corbett NP, Harike Bird Sanctuary, and Keoladeo NP. Nesting was recorded after 82 years in 1985 at Keoladeo NP, Bharatpur, Rajasthan (Prakash 1989) and its breeding biology was studied for the first time. In 2009–2010, three pairs bred in Keoladeo NP (Rishad Naoroji *pers. comm.* 2010). Rahmani (2012) has compiled all the major records from India. Here we give records from Uttar Pradesh.

Manoj Sharma (*pers. comm.* 2013) has recorded this species at **Okhla** Bird Sanctuary, on **Chambal** river that flows through Rajasthan, Uttar Pradesh, and Madhya Pradesh, and from areas in Uttar Pradesh bordering Bharatpur. It is likely to be more widespread in Uttar Pradesh than recorded till now.

Ecology: The Indian Spotted Eagle is mainly found in lightly wooded forest, forest clearings, cultivated areas, and even urban gardens. It is sometimes found near water, but is not as dependent on wetlands as the Greater Spotted Eagle. Its food consists of birds, small mammals, and amphibians. It nests on trees in open countryside. It occurs in low density, and difficulty in identifying the species compounds the problem of recording. The general ecology of this species was

TAXONOMY

It was described in 1831 as *Morphnus hastatus* by Lesson. Baker (1928) in *Fauna of British India* (Vol. 5, No. 75), called it *Aquila pomarina hastata*, the Small Indian Spotted Eagle. Ali & Ripley (1968) named it *Aquila pomarina hastata*, the Lesser Spotted Eagle. Two races were recognised: the nominate *pomarina* and the somewhat slimmer *hastata*. Many earlier ornithologists referred to the Indian race *hastata* variably as the Indian Spotted Eagle, Small Spotted Eagle, or the Long-legged Eagle, and treated it as specifically distinct from *pomarina* (Naoroji 2007). The *hastata* race was resident in India, while race *pomarina* was migratory and bred in Eurasia. Recently, Parry *et al.* (2002) have suggested separating the two races or subspecies into full species as there is enough morphological, ecological, and behavioural evidence, further proved by DNA sequencing (Vali 2006). Rasmussen & Anderton (2005) and Naoroji (2007) consider the Indian Spotted Eagle *Aquila hastata* a full species. The Lesser Spotted Eagle *Aquila pomarina* possibly does not occur in India (vagrant migrant to extreme northwest Pakistan, Afghanistan, according to Rasmussen & Anderton 2005). Naoroji (2007) has not included the Lesser Spotted Eagle in his book *Birds of Prey of the Indian Subcontinent*.

studied in detail at Keoladeo National Park (Prakash 1989) and in south India (Shivprakash *et al.* 2006). All existing knowledge of this species was compiled by Naoroji (2007).

Threats: Like all other raptors in India, it is also threatened by the destruction and disturbance of its habitat, modification of forest/grassland mosaic across southern India and Indochina, and also through pesticide poisoning. Research is required to know the specific threats to this species and how to mitigate them.

Conservation measures underway: It is protected under the Indian Wildlife (Protection) Act, 1972, not specifically, but under Eagles, thus it falls within the ambit of Schedule I of the Act. It is found in many well-known PAs/IBAs such as Keoladeo, Kaziranga, Corbett, Ranthambore, and Mudumalai TR (Naoroji 2007).

RECOMMENDATIONS

As it is a semi-endemic species in India, we have great responsibility for its protection. We give the following recommendations for research and conservation.

(1) Conduct proper surveys throughout the state to know its status. After this information is gathered, it would be possible to identify areas for protection and monitoring.

(2) Study its ecological requirements that include habitat, territory, prey base and prey availability, survival and mortality rates, and local movements outside the breeding season.

(3) The use of persistent organochlorine pesticides should be discouraged, and farmers should be encouraged to use organic pesticides.

(4) Study its movement and home range through satellite tracking.

Lesser Adjutant

Leptoptilos javanicus (Horsfield 1821)

DHRITIMAN MUKHERJEE

IUCN and BirdLife International (2013) place Lesser Adjutant in the Vulnerable category due to its small declining population, particularly as a result of hunting in some countries of its range. Its number is estimated to vary between 6,500 and 8,000.

Field Characters: It is the smallest member of the genus *Leptoptilos*, but still a big bird of 122–129 cm height, *c*. 5 kg (11 lb) weight, and 210 cm wingspan. It is dark grey-black above, white below, with naked head and neck, and a dirty yellowish wedge-shaped massive bill. It has sparse hair-like feathers on the naked head and neck, hence its old name Hair-crested Adjutant. Non-breeders have mostly yellowish head and neck skin with vinous-tinged sides and contrasting pale forehead. Breeding males show coppery spots on median coverts, narrow whitish edges to lower scapulars, tertials, and inner greater coverts, and redder sides to the head. Juvenile is duller and less glossy above, with more down on head and neck.

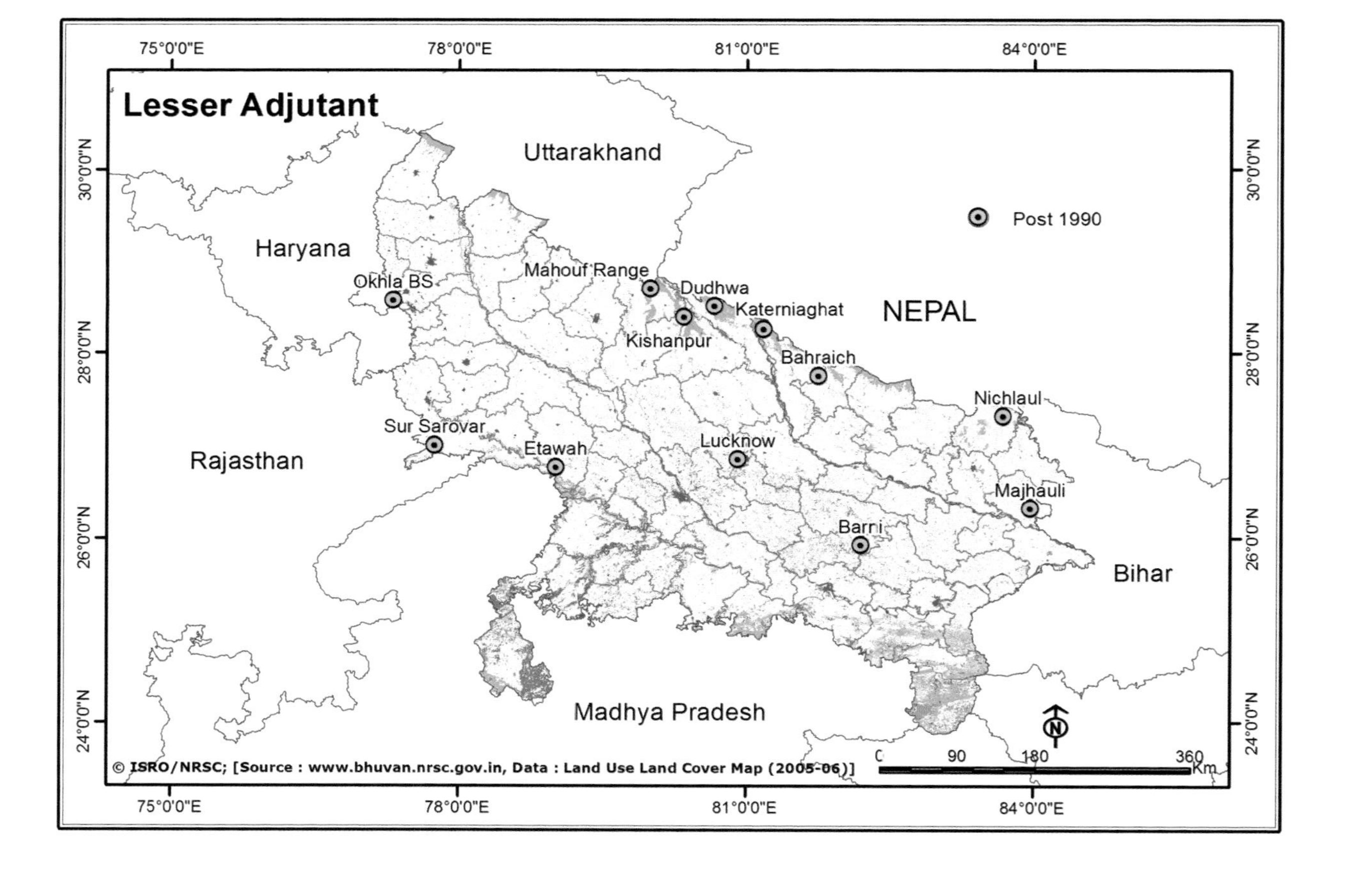
Lesser Adjutant
Uttarakhand
Haryana
Rajasthan
Madhya Pradesh
NEPAL
Bihar
Post 1990
Okhla BS
Mahouf Range
Dudhwa
Kishanpur
Katerniaghat
Bahraich
Nichlaul
Sur Sarovar
Etawah
Lucknow
Majhauli
Barri
75°0'0"E
78°0'0"E
81°0'0"E
84°0'0"E
30°0'0"N
28°0'0"N
26°0'0"N
24°0'0"N
0 90 180 360 Km
© ISRO/NRSC; [Source : www.bhuvan.nrsc.gov.in, Data : Land Use Land Cover Map (2005-06)]

Distribution: The Lesser Adjutant has an extensive range across South and Southeast Asia. It is found all over India, particularly in well-watered tracts. Rahmani (2012) has compiled recent records. Here we give records from Uttar Pradesh.

Records from Uttar Pradesh: This bird is recorded from most of the *terai* areas. Some records are as follows: six birds in grassland in **Katerniaghat** (April 2008); one bird in Sonaripur range in **Dudhwa** NP (March 2009), perched on an almost leafless tree; one bird in Singhrana Taal in **Sohagi Barwa** WLS foraging in a wetland infested with weeds (December 2009). Amit Mishra (*pers. comm.* 2013) spotted a flock of 12 birds in **Katerniaghat** WLS (December 2009) and six birds in Sathiana range of **Dudhwa** NP. Jaswant Singh Kalair (*pers. comm.* 2013) spotted 15 birds in Jhadi Taal in **Kishanpur** WLS in May 2010. A solitary bird was recorded in low lying roadside area on the **Etawah-Mainpuri** road in October 2012 (Neeraj Mishra *pers. comm.* 2013). More than 50 were sighted in Jhadi Taal in **Kishanpur** Wildlife Sanctuary in May 2012 and a group of 26 scattered over Bankey Taal in **Dudhwa** in June 2011.

During a bird survey of Sohelwa and Sohagi Barwa WLSs on November 20, 2013 by a BNHS team, three birds were seen near **Barni**, Siddharth Nagar district. One bird was recorded in a wet grassland in **Sohagi Barwa** Sanctuary on November 22, 2013. The BNHS team found up to ten birds in Belhiya Bazaar, near North Chowk, **Sohagi Barwa** WLS on December 19, 2013, four birds foraging in a wetland in South Chowk Range, **Sohagi Barwa** WLS on February 6, 2014, and one in Mahauf Range, **Pilibhit** Forest on February 20, 2014.

The Lesser Adjutant can be spotted in shallow areas of wetlands like **Sur Sarovar**, **Patna Jheel**, **Lakh-Bahosi**, and **Sandi** during the summer months of

ASAD R. RAHMANI

Lesser Adjutant has been found breeding in Dudhwa National Park

May to July looking for food, or perched atop dry tree stumps near waterbodies. In **Okhla** Bird Sanctuary, six were sighted on February 4, 1990 by Harris (2001), but since then there has been no sighting (Urfi 2003).

Its breeding has been confirmed only from **Dudhwa.** If a proper survey is done in Uttar Pradesh, perhaps more breeding areas will be detected.

Ecology: The Lesser Adjutant is an adaptable species found in forest pools, shallow open jheels, man-made wetlands, edges of reservoirs, drying roadside pools, and coastal wetlands, wherever it can get food. It nests on tall trees preferably in forests, but wherever it is not molested, as in many places in Assam, nests have been found on roadside avenue trees and even inside towns (e.g., Nagaon). Nesting is either in loose scattered colonies, sometimes up to eight nests found in a tree, but also solitarily (e.g., Dudhwa). Although it is solitary or seen in small scattered groups, sometimes 10 to 15 birds are seen in close proximity. For example, on May 22, 2010, Jaswant Singh Kalair (*pers. comm.*) photographed 15 Lesser Adjutant in Jhadi Taal of Kishanpur WLS.

Threats: The main threat to this species is destruction of wetlands and overfishing. Fortunately, hunting is not a major threat to this species in India as it is reasonably tolerated by people. It is considered not good to eat and unclean due to its feeding habits. Intensive use of pesticides in paddy fields is an indirect threat, as it results in loss of prey and biomagnification of pesticides in its body. Not much work has been done on this aspect. For instance, no sighting of Lesser Adjutant in Hastinapur WLS, Okhla Bird Sanctuary, and other existing habitats (such as Nawabganj) for almost two decades may be attributed to the above-mentioned threats. Not much work has been done on this aspect.

Conservation measures undertaken: The Lesser Adjutant is protected under the Indian Wildlife (Protection) Act, 1972. It is listed in Schedule IV of the Act. It is found in many PAs/IBAs of Uttar Pradesh.

RECOMMENDATIONS

(1) Periodic coordinated surveys should be carried out in the state.
(2) Protect nesting colonies outside PAs.
(3) Expand conservation awareness programmes.
(4) Research the species' use of and dependence upon agricultural areas including rice paddies.
(5) Develop wetland conservation policy.
(6) Strict control on use of zero-net (mosquito net) fishing and poisoning of wetlands.
(7) Study its movement through ringing/banding and satellite tracking.
(8) Develop protocol for yearly monitoring involving local communities and civil society.

Indian Skimmer

Rynchops albicollis Swainson 1838

DHRITIMAN MUKHERJEE

BirdLife International (2001, 2013) justifies the listing of the Indian Skimmer as Vulnerable, as its population is undergoing rapid decline due to widespread degradation and disturbance of its habitat in lowland rivers and lakes.

Field Characters: A large tern-like bird of 40–43 cm with a characteristic large orange bill, with the lower mandible elongated and the bill highly compressed laterally. The lower mandible has a yellow tip. Adult all black above, with white forehead and collar, and white below. It has long pointed wings, projecting much beyond its tail. Small, bright red legs characteristic. In flight, it displays a white trailing edge to the wing, and short forked tail with blackish central feathers. Sexes alike, though females are smaller. Non-breeders duller and browner above. Juvenile has dusky orange bill with blackish tip, paler brownish grey crown and nape with dark mottling, paler, more brownish grey mantle, and whitish to pale buff fringing scapulars and wing coverts. Call is a nasal *kap* or *kip*, particularly in flight and when disturbed.

Distribution: The Indian Skimmer is found on larger rivers from Pakistan through Nepal and India to Bangladesh and Myanmar. It was common in the 19th century in Myanmar, Laos, Cambodia, and Vietnam, but there are very few recent records from Myanmar and none from Laos, Cambodia, or Vietnam (BirdLife International 2001, 2013).

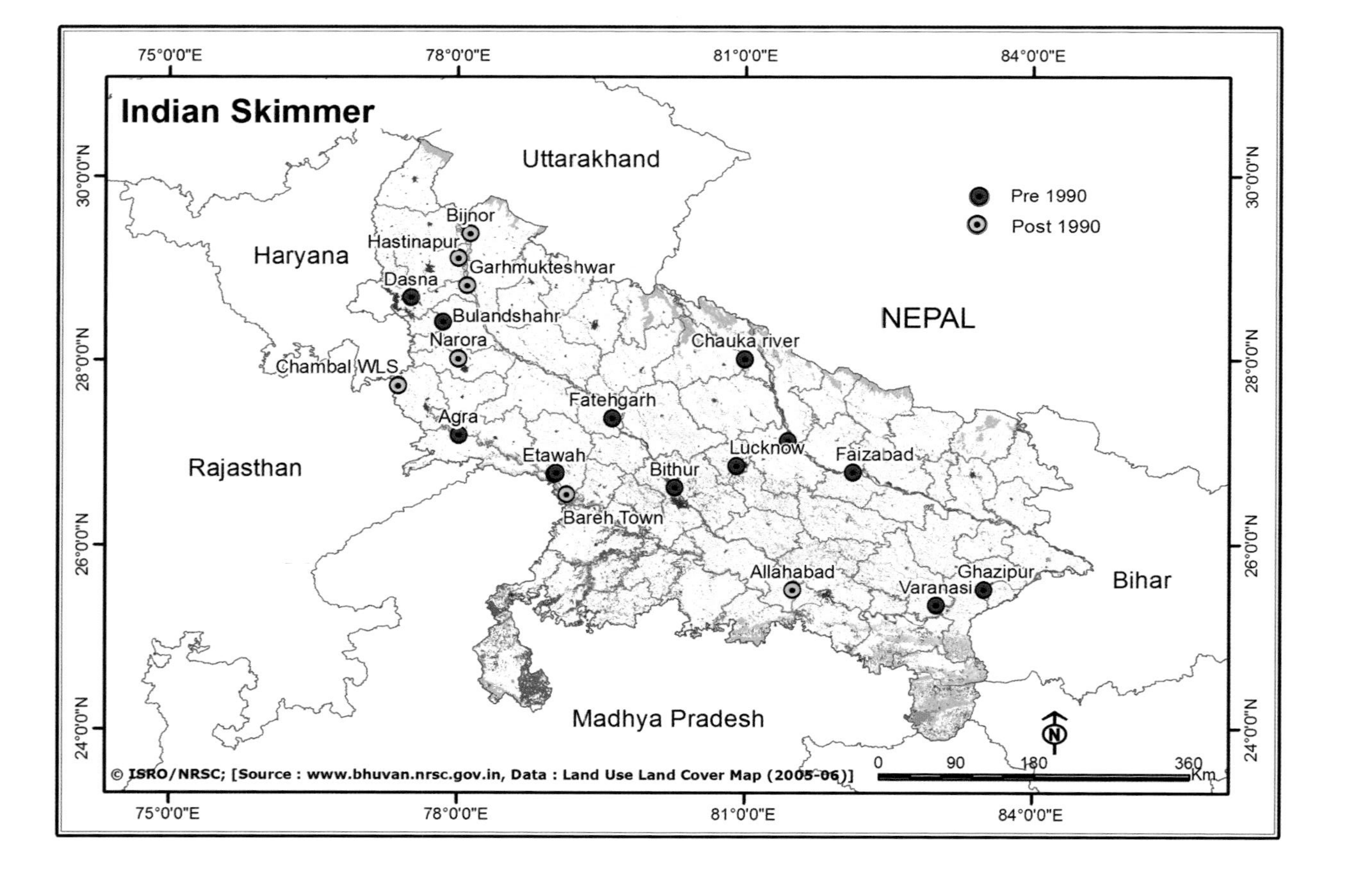

Indian Skimmer
Pre 1990
Post 1990
Uttarakhand
Haryana
NEPAL
Rajasthan
Madhya Pradesh
Bihar
Bijnor
Hastinapur
Garhmukteshwar
Dasna
Bulandshahr
Narora
Chambal WLS
Agra
Etawah
Bareh Town
Fatehgarh
Bithur
Chauka river
Lucknow
Faizabad
Allahabad
Varanasi
Ghazipur
0
90
180
360
Km
75°0'0"E
78°0'0"E
81°0'0"E
84°0'0"E
30°0'0"N
28°0'0"N
26°0'0"N
24°0'0"N
© ISRO/NRSC; [Source : www.bhuvan.nrsc.gov.in, Data : Land Use Land Cover Map (2005-06)]

In India, it is mainly found in the north, from Punjab (rare) through **Uttar Pradesh**, Madhya Pradesh, and Bihar to West Bengal, extending to Orissa (Chilika) and the Brahmaputra river. Possibly a separate population is harboured by Narmada, Mahanadi, Tapti, Godavari, and Krishna rivers in Andhra Pradesh and Orissa. Rahmani (2012) has compiled all the latest records from India. Here we give records of this species from Uttar Pradesh.

Records from Uttar Pradesh: The National Chambal Sanctuary on the borders of Uttar Pradesh, Rajasthan, and Madhya Pradesh is the most important sanctuary in India for this bird. As this river sanctuary is on the border of Rajasthan-Uttar Pradesh and Madhya Pradesh-Uttar Pradesh, we have included some records from these neighbouring states also. Sundar (2004) has conducted some studies in the **National Chambal** Sanctuary in Etawah district between December 1999 and February 2002 and conducted a census of waterbirds by boat (travelling at *c.* 20 km/h) between January 27 and February 5, 2002 along a stretch of 180 km. He counted 341 Indian Skimmer during five days of survey. Earlier, Sharma & Singh (1986) conducted a survey from December 1984 to February 1985 from **Pali** to **Pachnada** (425 km) and counted a total of 347 Indian Skimmer. During their second survey from December 1985 to February 1986, they observed 311 birds. Later, Sharma *et al.* (1995) reported 555 individuals in the **National Chambal** Sanctuary. A flock of 37 birds was seen at the **National Chambal** Sanctuary, on a small island sharing space with a female Gharial, 6 km upstream of Nadgawan. Sundar (2004) found 45–50 nests on an island near **Bareh** town (26° 31' 84" N, 79° 08' 26" E). The latest report (2010) is that in **National Chambal** Sanctuary, there are about 230 individuals left (R.K. Sharma *pers. comm.* 2010).

In a survey of Gharial in **Hastinapur** WLS, Sanjeev Kumar Yadav (*pers. comm.* 2010) saw 50 Indian Skimmer, on an island in front of **Jalalpur** village, Meerut district. Rajat Bhargava has recorded 15–20 birds of this species at **Bijnor** Barrage and in **Hastinapur** WLS in the winter of 2011–2012 (Bhargava 2012). The species has also been recorded at **Garhmukteshwar,** Hapur district in 2010 and **Allahabad** in 2013 at the Sangam confluence of Ganga and Yamuna rivers (Rajat Bhargava *pers. obs.* 2013). Information is available that shows the presence of this bird in **Patna** Bird Sanctuary, Jalesar *tehsil* in **Agra** district.

Another potential area with good presence of this bird is **Narora** in **Bulandshahr** district. Narora is an IBA and the lone Ramsar site in the state. During the Water Bird Count in the years 2012 and 2013, more than 50 and 70 birds were reported from this IBA (P.D. Mishra & Amit Mishra *pers. comm.* 2013). Flocks consisting of 15–30 individuals have been seen on different riverine islands on a 12 km stretch between **Narora Barrage** and **Karnabas** during different visits by the authors, despite much anthropogenic disturbances. Presence of this bird cannot be ruled out in identical riverside areas which suit its ecological and behavioural needs.

With such specific site preference, Indian Skimmer has been observed and studied only at National Chambal Sanctuary, where its occurrence and successful nesting has been reported. Gurmeet Singh (*pers. comm.* 2013) reported it from Bah Wildlife Range, **National Chambal Sanctuary**. He writes, "From February 2013 to June 2013, Gohra Island, Karkoli island, and Mau island were under observation. Main purpose was to keep vigil on the nesting sites and allow protection of nest from the predators. A total of 87 birds was seen and 37 nests with 111 eggs were recorded. Successful hatching from 108 eggs was also reported. Young birds took first flight with the first shower of monsoon and took off in the 1st week of June." In February 2013, Neeraj Mishra (*pers. comm.* 2013) also sighted 12 birds on the sand banks of River Chambal.

Ecology: The Indian Skimmer occurs primarily on larger, sandy, slow-moving, lowland rivers, around lakes and adjacent marshes, and in the non-breeding season on estuaries and coasts. It may hunt singly, but mostly in small flocks of 10–15 birds, sometimes more, much like a tern, skimming the water, the projecting tip of the lower mandible immersed in water at an acute angle, and as soon as its prey is struck, the mandibles are closed. The prey, generally small fish, is gulped head first. The Skimmer feeds mainly on fish but also takes small crustaceans and insect larvae (del Hoyo *et al*. 1996). It often feeds at dusk and through the night. It has a high, nasal, screaming call, but is often silent.

It breeds colonially on small sandy islands in larger rivers, and also on large, exposed sand-bars and islands, mostly from April onwards. There is no attempt at nest-making; a scrape on the sand serves as a nest where three or four eggs are incubated mostly by the female, with some assistance by the male. However, the males live near the nesting colony and help in chasing away small ground and aerial predators. Chicks are semi-precocial and start moving around within five to six days and are perfectly camouflaged.

Sundar (2004) found that at the National Chambal Sanctuary they nest on small islands along with Little Tern *Sterna albifrons* and Small Pratincole *Glareola lactea*. Nests of the three species were not interspersed, but instead formed distinct clumps, with only an occasional nest of one species found in the nesting clump of another. Sundar (2004) also found that the nests were spaced irregularly. While most nests were on high and dry parts of the island, three nests were located close to the water where the sand was wet.

Threats: The Indian Skimmer is under multiple sources of pressure. Although it is legally protected under the Indian Wildlife (Protection) Act, 1972, its habitat is under heavy anthropogenic pressures. Most of the large river islands in Ganga, Yamuna, Ghaghra, and Chambal are utilised by humans for summer cultivation of vegetables and watermelon, and most of these rivers are dammed to divert water for irrigation, drinking or industrial purposes, reducing some parts to narrow streams. Many eggs and chicks are destroyed by the sudden release of water from dams (e.g., Ghaghra river in Uttar Pradesh). Increasing human settlements near rivers and temporary settlements on islands bring with them crows, cats, and dogs, which exert additional predation pressure. Many old nesting colonies in the Ganga river have been abandoned. Pollution and spread of Water Hyacinth *Eichhornia crassipes* in lakes decreases the food supply.

The only well-known Indian breeding population in the National Chambal Sanctuary is threatened by increasing demand for drinking water to supply Dholpur, Bharatpur, and 999 villages in Rajasthan, lift irrigation and hydroelectric dams. Chambal water is already depleted due to construction of dams upstream in Rajasthan (Kota, Jawahar Sagar, Gandhi Sagar, Rana Pratap Sagar).

Conservation measures underway: It is not listed in the Indian Wildlife (Protection) Act, 1972. It occurs in some PAs/IBAs where poaching is under control. Some nesting sites benefit indirectly through protection of sand islands intended for more high-priority species such as Gharial and Mugger. A large stretch of Ganga river in Bulandshahr, Meerut, Hastinapur, and Moradabad is protected for Dolphin conservation. This stretch has been declared a Ramsar Site (Islam & Rahmani 2008).

RECOMMENDATIONS

(1) Start a long-term conservation management action plan which could be integrated into the Dolphin and Gharial action plans.

(2) Identify and protect all nesting colonies, particularly in Ganga, Ghaghra, Yamuna, Chambal, Son, and Ken rivers, and also the rivers in south India.

(3) Involve local people in the protection of nesting sites.

(4) Reduce the population of crows and stray dogs at its nesting colonies.

(5) Start an ecological study to identify the Indian Skimmer's specific breeding requirements and to see whether these can be provided artificially (e.g., raised platforms to prevent sudden flooding of nests by release of water from dams).

(6) Liaison with the irrigation department to control the flow of water.

(7) Campaign for strict protection of the National Chambal Wildlife Sanctuary, with particular reference to the maintenance of optimum water levels during the breeding season.

(8) Make Indian Skimmer an avian icon of river systems of India for tourists. However, no tourist should be allowed near nesting colonies.

(9) Study the impact of increased use of Ganga river as a waterway.

Great Slaty Woodpecker

Mulleripicus pulverulentus (Temminck 1826)

SIMON VAN DER MEULEN

In 2001, BirdLife International uplisted Great Slaty Woodpecker to Vulnerable as it has suffered a rapid population decline over the past 20 years (three generations) due to loss of primary forest cover throughout much of its range. However, the true rate of decline may be greater than currently estimated, and evidence of such a decline may result in the species being uplisted further.

Field Characters: Great Slaty Woodpecker is one of the largest woodpeckers in the world, slightly larger than the House Crow, *c.* 51 cm long. As the name indicates, it is overall slaty grey. Chin, throat, and foreneck are buffy yellow, and bill long and pale. The male has a short, broad, crimson moustachial stripe (Ali & Ripley 1987). Juvenile is brownish with a paler throat and indistinct white scaling on crown (Rasmussen & Anderton 2012).

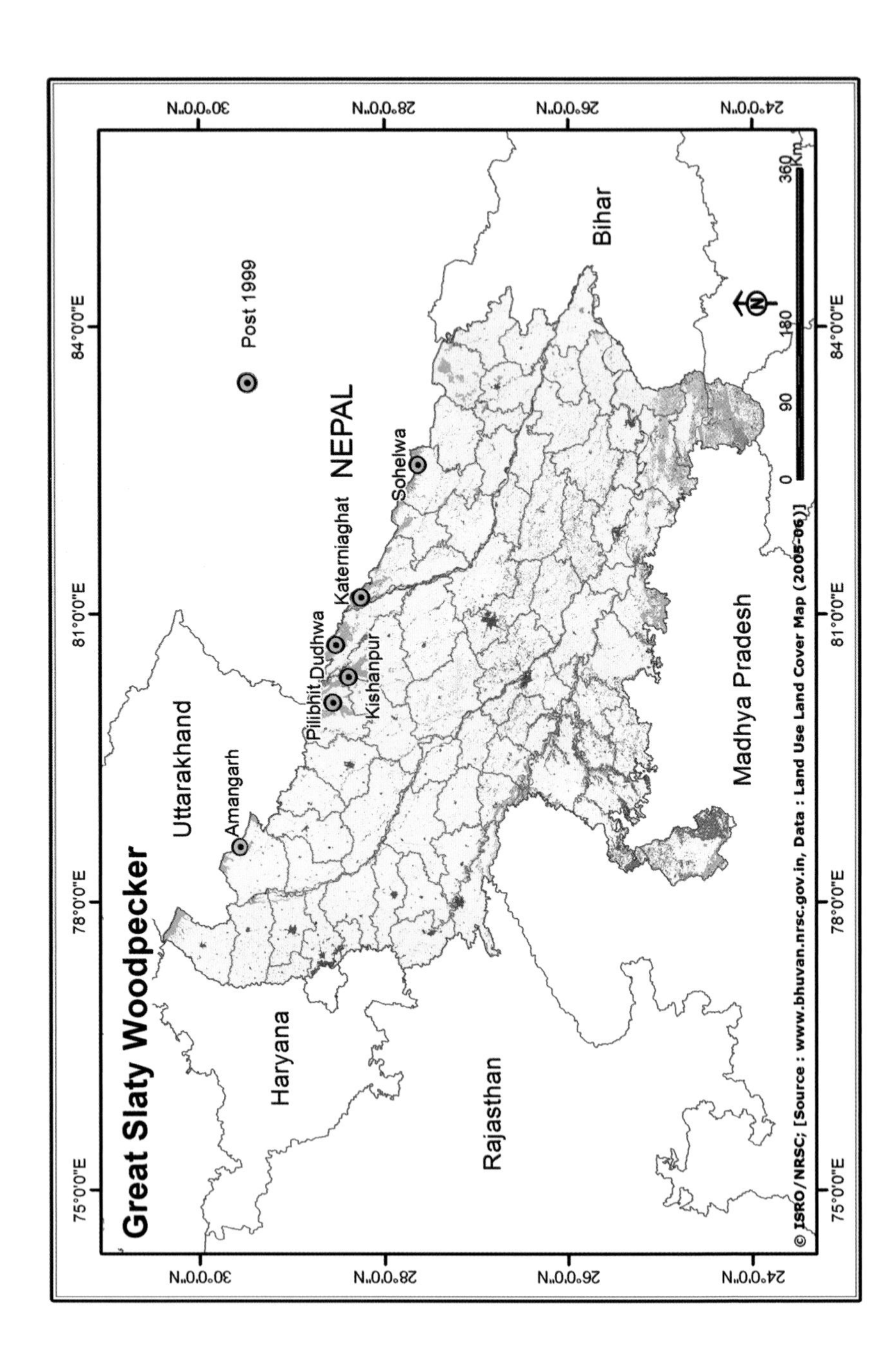
Great Slaty Woodpecker
Post 1999
NEPAL
Uttarakhand
Amangarh
Pilibhit
Dudhwa
Kishanpur
Katerniaghat
Sohelwa
Haryana
Rajasthan
Madhya Pradesh
Bihar
75°0'0"E
78°0'0"E
81°0'0"E
84°0'0"E
30°0'0"N
28°0'0"N
26°0'0"N
24°0'0"N
0
90
180
360
Km
© ISRO/NRSC; [Source : www.bhuvan.nrsc.gov.in, Data : Land Use Land Cover Map (2005-06)]

Distribution: This woodpecker has a wide distribution from India and Nepal foothills to southern China, Myanmar, Laos, Vietnam, Cambodia, and Thailand, and through peninsular Malaysia and Singapore to the western islands of Indonesia and the Philippines (BirdLife International 2013).

In India, it is found in Himachal Pradesh, Uttarakhand, **Uttar Pradesh**, West Bengal, Sikkim to Assam, Arunachal Pradesh, Nagaland, Manipur, and Mizoram. It occurs in climax semi-evergreen, evergreen, and moist deciduous forests, from the plains to *c*. 1,000 msl. In its northwestern distribution range, it is mostly associated with Sal *Shorea robusta* forests. It is uncommon everywhere, perhaps due to its specific habitat requirement of mature old trees. Based on the studies by Raman Kumar, Rahmani (2012) has given its distribution and status in north India.

Records from Uttar Pradesh: It is mainly found in the mature Sal forests of **Dudhwa** NP, **Katerniaghat** WLS, **Kishanpur** WLS, and **Pilibhit** RF. It was also reported earlier by Javed & Rahmani (1998) from **Dudhwa**. Fazlur Rahman (*in litt*. 2011) noted six together at a single Sal tree in April 2008. Many nest holes are found in May. Manoj Sharma (*pers. comm*. 2011) considers the extensive mature Sal forests of **Dudhwa** NP as one of the best places to see this bird. For example, in December 2008, Amit Mishra (*pers. comm.* 2010) spotted as many as five birds perched on a dead tree in Sathiana range of **Dudhwa** NP.

Rajat Bhargava and Nikhil Shinde of BNHS during their bird surveys of the Terai found three in the Rampur Range of **Suhelwa** WLS on December 29, 2013, and one in Dudhwa Range of **Dudhwa** NP on March 14, 2014. Sanjay Kumar has seen it in the Sal forest of **Amangarh** RF in Bijnor district in November 2013, and again in February 2014.

Ecology: It is a social bird, found in small parties of three to six in tall dense forest, and keeps constant contact with others by a loud, raucous call. It feeds woodpecker-like on trunks, moving up the trunk or branches in search of insect grubs. It flies from one patch of forest to another in follow-my-leader style through the tree-tops or high above the forest canopy (Ali & Ripley 1987). Nesting starts in spring (March) and extends up to late summer (July). Nest hole is excavated on the trunk of a large tree, sometimes very high up. Clutch size is three or four, incubation period not known. There is circumstantial evidence that the Great Slaty Woodpecker has cooperative breeding (involving more than two individuals), besides pair breeding. Lammertink (2004) found two nests of Great Slaty Woodpecker, one of which was attended by two males and one female. Raman Kumar & Ghazala Shahabuddin (*pers. comm*. 2011) have also confirmed cooperative breeding in this species.

Threats: The main threat to this large woodpecker of climax forest with mature trees is forest destruction and forest degradation. Although much reduced since the enactment of the Forest (Conservation) Act, 1980, degradation goes on, particularly outside PAs. Hunting, at least in Uttar Pradesh, is not the major threat.

The Great Slaty Woodpecker lives in climax semi-evergreen, evergreen, and moist deciduous forest. In UP, it is found mainly in mature Sal forest

In a study on woodpeckers in the sub-Himalayan dipterocarp Sal forests of Corbett NP and Ramnagar Forest Division in Uttarakhand, Kumar *et al.* (2011) found that the Great Slaty Woodpecker was almost absent in managed forests and teak plantations. Nearly all observations of the species were in stands of mature Sal trees with a diameter >60 cm (Raman Kumar *pers. comm.* 2011, Shahabuddin & Kumar 2011).

According to Abrar Ahmed (*pers. comm.* 2011), in the bird markets of northeast India this woodpecker appears infrequently. One or two hunted individuals may be brought by the local Nishi, Garo, and Cachar tribals. In other parts of India, many woodpeckers are hunted by certain tribes such as Gulgulawas or Kurmi-Baheliyas for meat in the entire distribution range of Great Slaty Woodpecker. The Great Slaty may also get accidentally caught in hanging nets put up for parakeets and green pigeons in the Sal forest.

Conservation measures underway: It is listed in the Wildlife (Protection) Act, 1972. It is found in many well-protected PAs/IBAs such as Dudhwa, Katerniaghat, and Kishanpur. It also occurs in many reserve forests of Uttar Pradesh. No special effort is being made to protect this species or its habitat in India.

RECOMMENDATIONS

BirdLife International (2013) has suggested many measures for the protection of Great Slaty Woodpecker in its whole range. Rahmani (2012) gave recommendations for its protection in India, and most of these suggestions are valid for Uttar Pradesh.

(1) Conduct surveys to find out its exact distribution and status in different parts of the state.
(2) Study its ecology, biology, and exact habitat requirements.
(3) Protect old growth mature forests.
(4) Provide special protection to nesting and roosting sites, wherever found.
(5) Study its density in undisturbed and disturbed forests, and develop mitigatory measures for regeneration of forests.

Bristled Grassbird

Chaetornis striata (Jerdon 1841)

ANAND ARYA

Like all grassland birds of the Indian subcontinent, the Bristled Grassbird is rapidly declining owing to loss and degradation of its grassland habitat, primarily through drainage and conversion to agriculture. It therefore qualifies as Vulnerable (BirdLife International 2013).

Field characters: The Bristled Grassbird is a large (20 cm), chiefly brown warbler with thick streaks on upperparts and fine streaks on lower throat, and unmarked buffy underparts. It has a relatively short, thick bill and very pale supercilium. It has a heavy, broad, rounded tail with whitish wash, and pale uppertail with distinct black central stripe. Male is larger (10%), with rufescent forecrown and dark bill in breeding season; female has browner forecrown and paler bill (Rasmussen & Anderton 2012). Female appears to be paler than the male and her bill is not so solidly black (Grewal 1996). The genus *Chaetornis* is characterised by unfeathered lores and five exceptionally strong rictal bristles arranged in a vertical row in front of the eyes (Baker 1924).

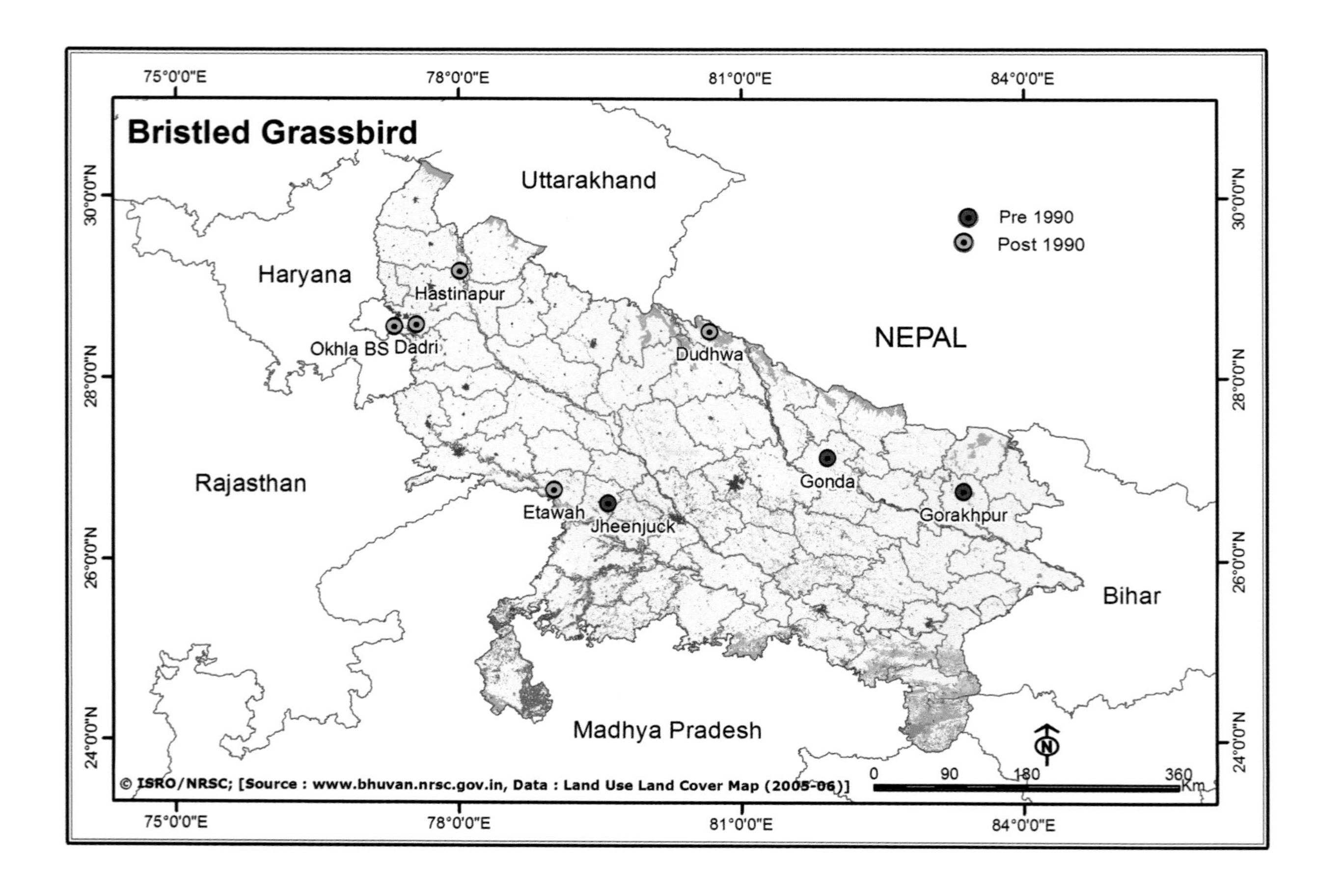
Bristled Grassbird
Pre 1990
Post 1990
Uttarakhand
Haryana
Hastinapur
Okhla BS
Dadri
Dudhwa
NEPAL
Rajasthan
Gonda
Etawah
Jheenjuck
Gorakhpur
Bihar
Madhya Pradesh
75°0'0"E
78°0'0"E
81°0'0"E
84°0'0"E
30°0'0"N
28°0'0"N
26°0'0"N
24°0'0"N
0
90
180
360
Km
© ISRO/NRSC; [Source : www.bhuvan.nrsc.gov.in, Data : Land Use Land Cover Map (2005-06)]

Distribution: It is endemic to the Indian subcontinent: Pakistan, India, Nepal, and Bangladesh. Although widely distributed, nowhere is it common as it depends on grasslands which are under tremendous pressure in India. Rahmani (2012) has given recent records from India. In **Uttar Pradesh**, it is (was) reported from the following areas: Dudhwa National Park, Gonda, Etawah, Gorakhpur, and Jheenjuck jheel (Kanpur). Some are historical records. The species has also been reported from Meerut district including **Hastinapur** Wildlife Sanctuary (Bhargava 2012).

Distribution and movement of the Bristled Grassbird is not clearly understood, but it appears to make nomadic local movements in response to the rainfall pattern, often appearing at sites for only a few months, then disappearing again. This species could be overlooked and much under-recorded, especially in its wintering range (Shashank Dalvi *pers. comm.* 2010).

Records from Uttar Pradesh: A male in breeding plumage was collected by Donahue (1967) from near **Okhla**, but subsequently there has been no record from the area, although it was visited by many good ornithologists. In 1996, the species was found at **Okhla**, a little above the barrage, on the Yamuna river in an area of extensive marsh and reeds created by the impoundment of water by the barrage (Grewal 1996). On August 4, one singing male was observed in an area of waterlogged grass and reeds along some borrow pits. On August 10, another male was noted, about 150 m from the first male, and by August 18, three pairs were seen — the two original males seeming to have acquired mates. Unfortunately, in 1997 the whole area was bulldozed and all the grassland completely destroyed to "rid the area of mosquitoes". However, it has reappeared due to development of reed beds, as Manoj Sharma (*pers. comm.* 2013) recorded this bird from **Okhla** Bird Sanctuary in July 2012. In recent years, it was regularly noted from a small patch in **Dadri** wetlands near Noida/Delhi in summer (June–August, 2010), the birds can be seen singing from high perches and in the air (Anand Arya *pers. comm.* 2010).

Ecology: The Bristled Grassbird lives in tall grasslands and reeds, with some shrubs, preferably near water, but sometimes in dry grasslands also. In Nepal, it occurs in relatively open, short grasslands, mostly on dry soil, but also in moist areas with tall reeds and scattered bushes (BirdLife International 2013). In Corbett, Sharma (2007) found it in tall, dry, riverine grasslands with an average grass height of 2.5 m, with some reeds as high at 4 m.

It lives singly or in pairs, is very shy and skulks, but during the breeding season the males become very vocal and conspicuous, and cannot be missed in the right habitat. The territorial male sings from an exposed perch on a tall grass stem or a bush standing amongst the reeds, or in frequent sustained song flights. The song is a disyllabic *trew-treeuw* and *treeuw-trup*, sometimes alternated, repeated monotonously at 2–3 second intervals over a period of 10–15 minutes (Grewal 1996). The call is fairly sharp and loud, and carries over a considerable distance. Sharma (2007) found the song loud, a somewhat metallic *chwee-chew*, repeated monotonously every two seconds.

The breeding season is generally from May to September (Ali & Ripley 1987). In Corbett, Sharma (2007) found singing territorial males in June. These birds seem to have a few favourite song-perches in their territories, always tall reeds well above the average height of the grass. The males frequently fly above 40–50 m high in the air in wide circles of 250–300 m diameter while calling (Sharma 2007). The nest is made in dense grass or shrub, a ball-shaped bundle of grass with side entrance, where the female lays 4–5 eggs and apparently incubates alone.

Threats: Habitat destruction is the biggest threat to this species, as grassland protection is totally neglected in India. Most of the wet grasslands have been converted to sugar cane fields, particularly in the Terai. It is not known how the bird adapts to this changing scenario. Overgrazing, grass cutting, burning and commercial plantations are other major threats. Burning of grassland during the breeding period of this species is a major management issue in certain areas.

From May to September, there is frequent trapping of munias during their night roost in sugar cane fields and tall wet grasslands. Munias roost communally with weaverbirds, grassbirds, and babblers. Extensive trapping of munias, especially in wet tall grasslands shared with the Bristled Grassbird often results in the accidental trapping of the latter. At times these birds are withheld from capture sites to be retailed the next morning for release-bird trade, along with prinias and warblers. As these birds are neither pet nor meat birds, they are not fed and hence have very little chance of survival in urban habitation where they are bought by religious-minded people to be released (Ahmed 2000, 2012).

Conservation measures underway: Although this species is not listed by name in the Indian Wildlife (Protection) Act, 1972, shooting and trapping of all wild bird species is prohibited under the Act. Thus it is also protected. It is recorded from many PAs/IBAs (Islam & Rahmani 2004).

RECOMMENDATIONS

BirdLife International (2013) has suggested measures for all the range countries, and Rahmani (2012) has given recommendations for India. Many of these recommendations are valid for Uttar Pradesh.

(a) Study the ecology, behaviour, and habitat requirement of this species, preferably through colour marking (ringing).

(b) Formulate the best grassland burning regime practices to benefit this species.

(c) Conduct extensive all-India field surveys, especially in its winter range.

(d) Study the breeding biology of the species to arrive at specific strategies for conservation of their breeding sites.

(e) Identify potential breeding sites for this species and lobby through local NGOs to get these areas secured.

(f) Protect small wetlands, such as those near Dadri.

White-throated or Hodgson's Bushchat

Saxicola insignis Gray 1846

NIKHIL DEVASAR

The White-throated Bushchat or Hodgson's Bushchat is poorly known and has a small, declining population as a result of loss of its wintering grassland habitats to drainage and conversion to agriculture, overgrazing, and flooding. This justifies placing it in the Vulnerable category (BirdLife International 2013).

Field Characters: Among the 11 *Saxicola* species and subspecies found in the Indian subcontinent, generally varying from 13 to 15 cm in size, this is the largest (17 cm). Adult male has a snowy white throat, extending to form an almost complete white collar. It has black wings with much white on the coverts, chestnut breast grading into buff flanks and whiter centre of belly and vent. The adult male has a black head, ear-coverts, and mantle with rufous-brown fringes. The worn breeding male has a black head, nearly complete white collar, and mainly black upperparts with broad white rump (Rasmussen & Anderton 2005). Female

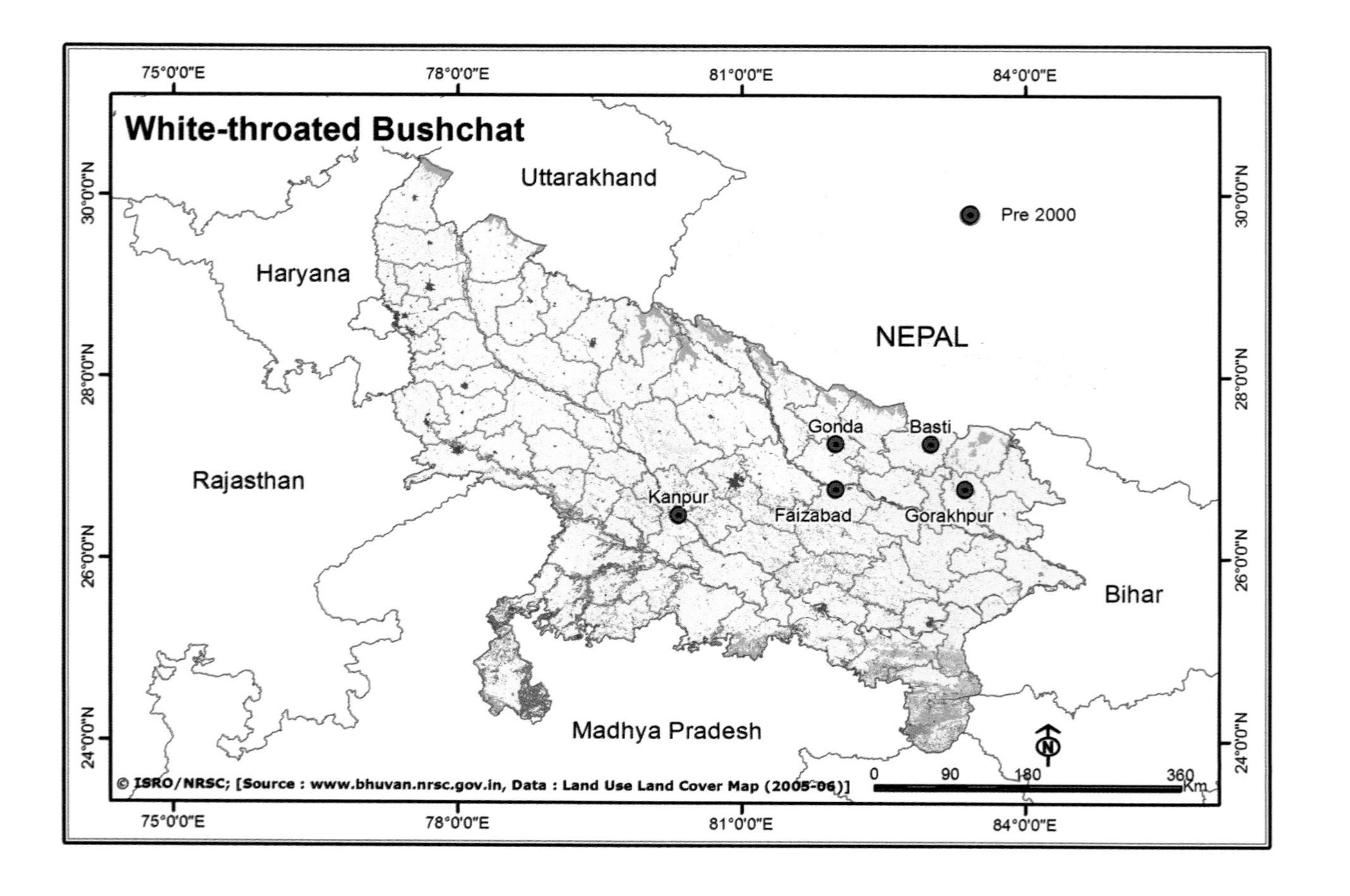

White-throated Bushchat
Uttarakhand
Pre 2000
Haryana
NEPAL
Gonda
Basti
Rajasthan
Kanpur
Faizabad
Gorakhpur
Bihar
Madhya Pradesh
0 90 180 360 Km
© ISRO/NRSC; [Source : www.bhuvan.nrsc.gov.in, Data : Land Use Land Cover Map (2005-06)]
75°0'0"E
78°0'0"E
81°0'0"E
84°0'0"E
30°0'0"N
28°0'0"N
26°0'0"N
24°0'0"N

has buffish supercilium, pale throat, and buffish crown and mantle. For more details of plumage variation in summer and winter and between age/sex classes, see Urquhart (2002) and Rasmussen & Anderton (2005).

Distribution: White-throated or Hodgson's Bushchat has been recorded from six countries only. Its breeding quarters lie in Russia, the mountains of Mongolia, and adjacent parts of Kazakhstan. It passes through northern and western China and Tibet, and winters in the Terai in Nepal and India, particularly in the Terai of **Uttar Pradesh** and Bihar, extending eastwards up to the Jalpaiguri *duar*s and western Assam. Rahmani (2012) has collated all the recent records from India. There are very few recent records from **Uttar Pradesh**. Manoj Sharma (*pers. comm.* 2013), who has done extensive birdwatching in Uttar Pradesh during the last ten years, has never seen this species in the state, which further proves its rarity. However, as it is found in the Nepal Terai, it is likely to occur in the UP Terai also. For example, Inskipp & Inskipp (1991) reported a maximum of 10 individuals from Kosi Barrage on March 17 and 18, 1982. A total of 29 individuals were recorded in December 1997 in Chitwan and Sukla Phanta in Nepal by Baral (1999). He estimates that a maximum of 110 individuals are likely to winter in Nepal, and recommends that its status be uplisted from Vulnerable to Endangered (Baral 1999).

Historical records are from "Cawnpore [Kanpur] to the Bhutan Doars" (Baker 1924). BirdLife International (2001) has mentioned records from Faizabad, Gorakhpur, Gonda and Basti (northern part divided into Siddharth Nagar). It is likely to occur even now in the floodplains of the Ganga with vast *sarpat* grasslands that exist in some areas. Such areas are generally not visited by many ornithologists, hence it is likely to remain undetected.

Ecology: It breeds locally in the subalpine and alpine parts of western and central Mongolia and in adjacent areas of Russia and possibly Kazakhstan (Urquhart 2002). It migrates through China and Tibet and winters in the Terai in Nepal and India, the Gangetic plains, and the *duar*s of northeastern India, generally below 250 msl. On migration, it occurs up to 4,500 msl (Tibet and China).

Urquhart (2002) and del Hoyo *et al.* (2005) have summarised the known facts about this species. It is extremely sporadic in its breeding range and absent from apparently suitable habitats. In its winter quarters also, it is patchily distributed or absent in likely habitats, albeit it is present in some well-protected areas (Corbett, Manas, Sukla Phanta, Chitwan). In Mongolia, it was found breeding between 2,800 and 3,100 msl. "Some pairs were relatively densely located with 3–5 pairs carrying food to their young in nests up to only a few tens of metres apart, whereas others were up to 1–2 km apart" (Urquhart 2002, p. 109). Nest is a bulky, thick-walled cup of grass, lined with wool, soft feathers, and dry moss, placed in an earth wall or rock crevice in shallow ravines or gullies sheltered by overhanging turf (del Hoyo *et al.* 2005). Clutch size is four or five, incubation apparently by female only, but both parents feed the chicks. The call is a metallic *teck-teck*. Its food consists of small insects, mainly beetles, insect larvae, and green vegetable matter. Baral (1999) reported an individual catching a flying moth.

In winter, it occurs in grasslands, sometimes quite disturbed ones also, and in a mosaic of grasslands and sugar cane fields. In Manas, Narayan & Rosalind (1997) observed it in vast open grasslands dominated by *Saccharum narenga* and *Imperata cylindrica*, with scattered bushes and trees. In the same area, the Collared Bushchat *Saxicola torquata* is also found. Narayan & Rosalind (1997) found that the White-throated Bushchat kept singly or in loose pairs, perching on the top of low (*c.* 75–150 cm) grass or shrubs, and often descending to the ground to feed. The same patch of grassland was occupied for days and they defended these small territories by chasing away any intruding *S. torquata*.

Threats: It appears to be a naturally rare species, with decline in numbers in recent years due to massive destruction of its wintering habitat in India and Nepal. However, if it can live in slightly disturbed areas and cane fields, it should be doing well as the natural Elephant Grass areas have been replaced by a mosaic of sugar cane fields, wheat fields, and grazing grounds. Good natural grasslands still survive in many protected areas such as Dudhwa, Kishanpur, Pilibhit, and Lagga-Bagga in Uttar Pradesh. There appears to be no major threat to the habitat in its breeding areas. It is not specifically targeted by bird traders as it is difficult to keep in captivity.

Conservation measures underway: Although it is not listed by name in the Indian Wildlife (Protection) Act, 1972, shooting and trapping of all wild bird species is prohibited under the Act. It is also in the CMS Appendix II. Small numbers are seen regularly in several protected areas (see Islam & Rahmani 2004).

RECOMMENDATIONS

(1) Conduct a thorough survey in the Gangetic plains and river islands still covered by Elephant Grass.
(2) Based on the survey results, conduct detailed studies on its ecological requirements in its winter quarters.
(3) Monitor populations, preferably through ringing and colour banding.
(4) Develop suitable grassland management protocol to benefit this species.
(5) Extend, upgrade, and link (where possible) existing protected areas, and establish new ones, in order to adequately conserve remaining tracts of natural grassland. Promote grassland regeneration.
(6) If necessary, control livestock grazing in some areas, at least in winter.
(7) Launch an environmental education campaign through appropriate publicity material to monitor its populations, involving local birdwatchers.

Yellow Weaver

Ploceus megarhynchus Hume 1869

BirdLife International (2013) justifies listing the Yellow Weaver as Vulnerable as this species has a small, rapidly declining, and severely fragmented population due to the loss and degradation of grasslands in the Terai, principally through conversion to agriculture and overgrazing. However, in view of its very small declining population and increasing threats, we suggest that it should be uplisted to the Endangered category.

Field Characters: A large (17 cm), large-billed, heavy-legged, and long-tailed weaver, with yellow rump, uppertail-coverts, head, and underparts, and dark ear-coverts. It has heavily streaked mantle, back, and scapulars. Female duller with paler, more buff-tinged yellow parts, particularly crown and nape. Female and non-breeding male Baya Weaver *Ploceus philippinus* look similar to Yellow Weaver, but they are smaller with shorter, narrower bill, and lack dark lateral

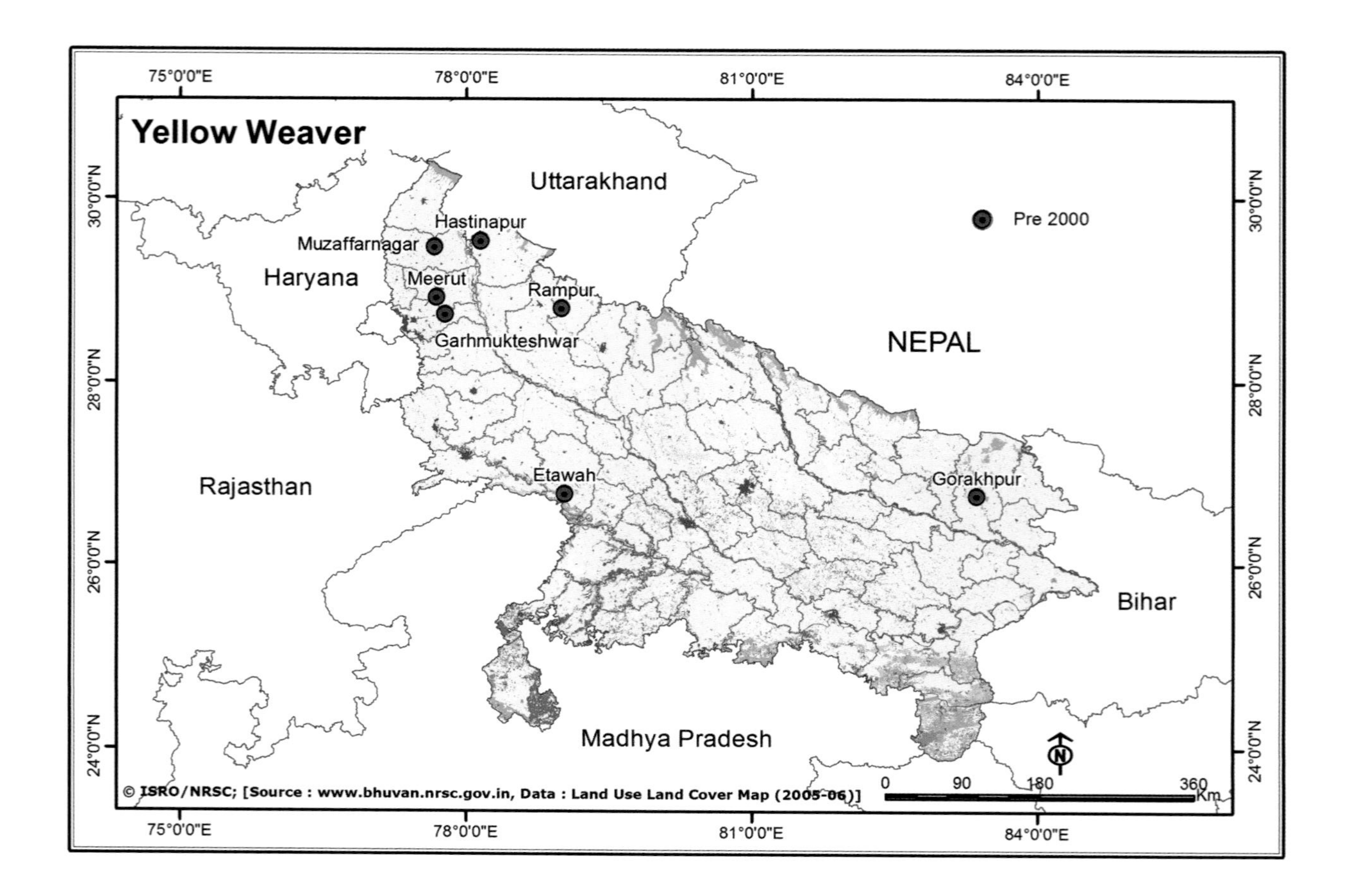
Yellow Weaver
Pre 2000
NEPAL
Uttarakhand
Haryana
Rajasthan
Madhya Pradesh
Bihar
Muzaffarnagar
Hastinapur
Meerut
Garhmukteshwar
Rampur
Etawah
Gorakhpur
30°0'0"N
28°0'0"N
26°0'0"N
24°0'0"N
75°0'0"E
78°0'0"E
81°0'0"E
84°0'0"E
0 90 180 360 Km
© ISRO/NRSC; [Source : www.bhuvan.nrsc.gov.in, Data : Land Use Land Cover Map (2005-06)]

breast-patch. For detailed descriptions of age/sex related plumage differences, see Ali & Ripley (1974) and Rasmussen & Anderton (2012).

P.M. LAD

Breeding male of the Yellow Weaver has blackish bill, reddish yellow head, bright yellow breast and belly, and dark brown cheeks

Distribution: The Yellow Weaver or Finn's Baya was considered endemic to India, but in May 1996 a small population was discovered in Sukla Phanta in Nepal by Baral (1998), about 100 km east of Kaladhungi in Uttarakhand where Ali & Crook (1959) rediscovered this bird. It is now known to have disjunct populations in the Terai and from eastern Nepal to Assam. It has always been very locally distributed, and the disappearance of several colonies in recent decades Indicates that it is declining. The recently discovered population in Nepal is estimated at <50 birds and its occurrence is erratic (BirdLife International 2013). Twelve were seen in two different flocks in the eastern part of Sukla Phanta grassland in Nepal on May 8, 1996 by Baral (1988). For further details see Giri & Choudhury (1996).

After two fruitless searches by Sálim Ali in 1934 (Ali 1935) and again in 1954 with H. Alexander, a colony of Finn's Weaver or Yellow Weaver was rediscovered in 1959 in Kumaon Terai (Ali & Crook 1959), which is now in Uttarakhand. In the present **Uttar Pradesh** (excluding parts that are now in the state of Uttarakhand), the species has been reported from very few areas in the past, with no recent records.

From the available published records, the species was reported from the **Hastinapur** Wildlife Sanctuary (Rai 1979) where a small population may still exist (Bhargava 2012). According to Bhargava (2000) it was reported from **Etawah**, **Garhmukteshwar** (Hapur), *khadar* areas of **Muzaffarnagar** district, Landour area near **Roorkee**, **Rampur**, and Kila near Parikshitnagar (**Meerut**), but there are no recent records from trappers or surveys carried out in 2002–2003 (Bhargava 2004) or after that. Some traditional bird trappers reported Yellow Weaver from **Gorakhpur** (*pers. comm.* to Rajat Bhargava 2002). Though this bird has not been reported by any birdwatchers or ornithologists in present day Uttar Pradesh, there is a likelihood of its presence, considering its erratic distribution.

Ecology: The ecology and behaviour of Yellow Weaver were studied by Ali & Crook (1959), Ambedkar (1968), and recently by Bhargava (2000). It is a gregarious

Yellow Weaver makes its nest on top of large trees and removes the leaves as shown above and below

species found in pure *terai* country with marshes and stands of *Imperata arundinacea* and *Saccharum spontaneum* with scattered isolated trees, particularly Silk Cotton (Semal), occasionally interspersed with patches of crop fields.

The Yellow Weaver forages in flocks and feeds on rice grains, Hemp *Cannabis sativa*, millets *Panicum* sp., and small insects. Bhargava (2000) found that on several occasions the birds fed on insects on the millet spike rather than on millet grains. The females were repeatedly seen returning to the nest with insects. It is some-times seen foraging on fallen paddy on village tracks.

Breeding occurs from May to September in colonies. In the non-breeding season, the bird may move around but comes back to traditional nesting sites every year. The nest is a large, globular, untidy structure, firmly woven with long strips of coarse grass, with the entrance on top, to one side. Interestingly, multiple nests built by single males are usually linked with connecting walls or by separate strands of vegetation bound firmly at each end to different individual nests (Ali & Crook 1959, Ambedkar 1968). Ambedkar (1968) mentions that Finn's Weaver has two distinct breeding periods: the first from May to the middle of July, and the second in August and September. He noted that in the first period the birds build nests in treetops, and in the second low down among *Typha* beds standing in water. They line the entire inside of the nest, unlike other weavers which line only the floor of the nest. Sometimes they build their nests near a nest of Black Drongo *Dicrurus macrocerus* to use the Drongo's pugnacious behaviour towards predators to their own advantage. Nesting colonies with no Drongo nest are often preyed upon by crows (Bhargava 2000). First year males, or late nesters, may make nests on tall reeds. Interestingly, Choudhury (2000) found them nesting on short Silk Cotton trees in Orang NP and shrubs in Kaziranga NP. Similar nesting behaviour was also recorded during a recent survey in 2013 by Rajat Bhargava, where three nests were found on short Silk Cotton trees near Pantnagar.

The sex ratio is two to four females per adult breeding male. The male probably practices sequential polygamy where a male mates with a number of females one by one. Once a female has selected a nest and started laying eggs, he goes to another female. According to Rajat Bhargava, research is needed on marked birds to further confirm this. Nest is built by the male, but incubation and chicks are reared by female only. Call is *twit-twit* but harsher than Baya Weaver, and the male's song is a subdued *twit-twit-tit-t-t-t-t-t-trrrrr wheeze whee wee we*. Other notes have also been recorded such as *skeer, skeer, skeer* or *tseer, tseer,* in aggression, a high pitched alarm call, and twittering on take-off or landing (Rasmussen & Anderton 2012).

Threats: The main threats are loss and modification of the Terai habitat, and illegal trapping for live-bird trade (Ahmed *et al*. 1996; Ahmed 1997, 1999). During the last 60 years, the Terai region has been almost totally converted to human-dominated landscape with agricultural farms, orchards, factories, canals, roads, expanding villages and cities, and very rapid human population growth. The Yellow Weaver's survival was dependent on large undisturbed uncleared tracts of tall Narkul *Arundo* spp. and Patera *Typha angustata* grass, which were historically present in several wetland pockets of Meerut (especially Hastinapur WLS), Ghaziabad, Rampur, and Hapur districts. The clearance of these tall wet grasslands, cutting of Semal trees and monoculture plantation, lack of available microhabitat ultimately led to the disappearance of this species from most of its former known range in Uttar Pradesh. This threat, combined with the rising population of crows (*Corvus splendens* and *C. macrorhynchos*), related to garbage and human habitations, was a direct blow that led to the systematic wiping out of entire nesting colonies, since crows preyed on the eggs and chicks at every possible opportunity.

It has been recorded in bird trade since 1901, and investigations on bird trade by Ahmed (1997, 1999) suggest that although not targeted by local trappers, the species may occasionally get caught at roost sites in sugar cane and tall grass which it shares with other species of weavers and munias.

Conservation measures underway: It is protected in India, and trapping and trade of the species has been banned since 1991, under the Wildlife (Protection) Act, 1972. Although it is not listed by name in the Act, all members of Ploceidae are listed in Schedule IV of the Act. It has been recorded from many PAs/IBAs (Rahmani 2012) such as Hastinapur WLS, Manas, Kaziranga, Orang, Dibru-Saikhowa, and Jaldapara, and Sukla Phanta Wildlife Reserve in Nepal.

RECOMMENDATIONS

(1) There is urgent need for a systematic survey of every known habitat of its former distribution in UP in various seasons to record the species' presence or absence, to work out a conservation plan for its *in situ* management.

(2) The western population is being wiped out mainly due to raids by crows on breeding colonies, and increasing loss of microhabitats required for breeding. The prime population has fallen so low and is so severely fragmented that it is unable to cope with the increasing crow population. Either crows should be eliminated or controlled, or a small population of Yellow Weaver should be reintroduced on an experimental basis in places such as Hastinapur Wildlife Sanctuary where necessary microhabitat for breeding is available. Due to the continued efforts of UP Forest Department and WWF-India in the last five years towards reintroduction of Gharials and increased patrolling, the species can indirectly benefit in Hastinapur WLS.

(3) In order to conserve this species and bolster its wild population, a conservation breeding project should be initiated at Hastinapur WLS or in the Moradabad-Rampur belt before the remaining unrecorded population gets extinct. The species is an ideal candidate for conservation breeding, since a crow-free environment is impossible to achieve in India (Bhargava 2000), or to have a weaver population large enough to mob out the crows during nesting.

(4) The species should be uplisted to Endangered as its global population may be less than 2000 mature individuals (Bhargava 2000, Rajat Bhargava *in litt.* 2013) and is rapidly declining.

(5) Detailed studies should be made all over its range in India every two years to monitor its status.

(6) Large Semal trees present where nesting is regularly observed should be identified and declared protected trees which cannot be felled. As many of them grow in agricultural landscapes and are often in private ownership, they are vulnerable to felling for agricultural and urban development.

Falcated Duck
Anas falcata Georgi 1775

DHRITIMAN MUKHERJEE

In India, the Falcated Duck is a rare migratory species. IUCN and BirdLife International (2013) have listed it as Near Threatened owing to a rapid decline in its population in China and very high levels of hunting, although elsewhere it is more abundant than was once believed.

Field Characters: Ali & Ripley (1987) describe it as follows: Male (breeding) strikingly peculiar and beautiful, with metallic bronzy green head with a chestnut-purplish crown and a bushy mane-like nuchal crest which falls over the hindneck and rests on the back, giving the appearance of a thick neck. Throat and foreneck white, with a narrow green collar near base. Its body plumage is mainly grey, wavily pencilled with black, the marking becoming bolder and more crescentic on the breast. Speculum glossy black and green, bordered in front by a grey band (wing-coverts); inner secondaries very long, sickle-shaped (falcated, hence its name), velvety black, white and grey, covering hind part of body and tail. Uppertail-coverts black, overtopping tail. Female quite drab, with greyish head, dark spotting and scalloping on brown underparts, and greyish-white fringes to exposed tertials. It is between 48 and 54 cm, about the size of a Gadwall *Anas strepera*. The eclipse male is like the female, but darker on the back and head. In flight, both sexes show a pale grey underwing.

Distribution: The Falcated Duck breeds from southeast **Siberia** and **Mongolia**, to the **Kuril Island** and northern **Japan** and **China**. Although the global population was previously estimated to be 35,000 individuals, recent counts indicate that it is considerably higher, totalling perhaps as many as 89,000 (BirdLife International 2013). In **India**, the Falcated Duck is found mainly in the northern parts as an uncommon winter migrant. Rahmani & Islam (2008) and Rahmani (2012) have given detailed site records of India. In Uttar Pradesh, historically, it has been reported from **Lucknow** and **Roorkee.** It has been reported from **Dudhwa** NP as an occasional migrant by Javed & Rahmani (1998) during their studies between 1991 and 1994.

Ecology: In India, not much is reported about its ecology except that it occurs singly or in pairs with other dabbling ducks on jheels and shallow waterbodies. In flight it makes a swishing sound, so characteristic of the Common Teal *Anas crecca*. In its breeding areas, it is usually seen in pairs or small parties. According to del Hoyo *et al.* (1992), it breeds in freshwater lakes, rivers, ponds, lagoons, and often in wooded country, while in the winter it can be seen on the coast and in larger, shallow waterbodies, in rice fields and flooded meadows. It feeds on seeds including rice and other grains, grasses, and also molluscs and insects. It starts breeding in May–June in single pairs or loose groups, nesting on the ground in vegetation near water. It lays 6–9 eggs and incubates them for 24–26 days (del Hoyo *et al.* 1992). Like other species of *Anas*, it is mainly vegetarian, and primarily feeds on emergent and submerged vegetation by dabbling and upending, but it also grazes on wet grasslands and crops. The male Falcated Duck has a clear low whistle, whereas the female has a gruff *quack*.

Threats: Degradation and drainage of its wetland habitat is the main threat in India, but elsewhere, BirdLife International (2013) found that hunting for food for subsistence and local markets is probably the major threat.

Conservation measures underway: Like all Anatidae species, it is protected under the Indian Wildlife (Protection) Act, 1972. It is listed in Schedule IV of the Act. It also occurs in some PAs/IBAs.

RECOMMENDATIONS

As it is an uncommon winter visitor in Uttar Pradesh, no specific recommendations can be made. In general, wetlands should be protected, and hunting and poisoning of all waterfowl should be stopped at once.

Ferruginous Duck

Aythya nyroca (Güldenstädt 1770)

DHRITIMAN MUKHERJEE

Based on the information gathered by BirdLife International (2013), IUCN has categorised the Ferruginous Duck as Near Threatened. There has been a rapid decline in its population in Europe, but evidence of decline in the larger Asian populations is sparse, and sometimes contradictory, so it currently remains listed as Near Threatened. Evidence of rapid decline in Asia would qualify the species for uplisting to Vulnerable (BirdLife International 2013).

Field Characters: An overall dark chestnut diving duck, *c.* 41 cm, with a large oval white patch on the belly (clearly visible in flight), with conspicuous white eyes in male visible clearly at short distance. Female is like the male but duller, with brown eyes. Both sexes slightly darker on back. Juvenile similar to adult, but belly and undertail are grey-buff. In flight, a broad white wing-bar extends to outer primaries.

Distribution: The Ferruginous Duck or White-eyed Pochard is widely distributed in the Palaearctic region from western Europe to western Mongolia. There is an isolated breeding population in **Libya**. In north **India**, it is a common winter migrant mainly to the Northeast, with scattered records from northern and southern India (Ali & Ripley 1987, Grimmett *et al.* 1999). Rahmani & Islam (2008) and Rahmani (2012) have given historical and recent records from India. Here we mention only records from **Uttar Pradesh**.

Records from Uttar Pradesh: The Ferruginous Duck may not be as uncommon in northern India as reported by Ali & Ripley (1987). Based on the information given by IBA workshop participants and our own surveys, it is also reported from

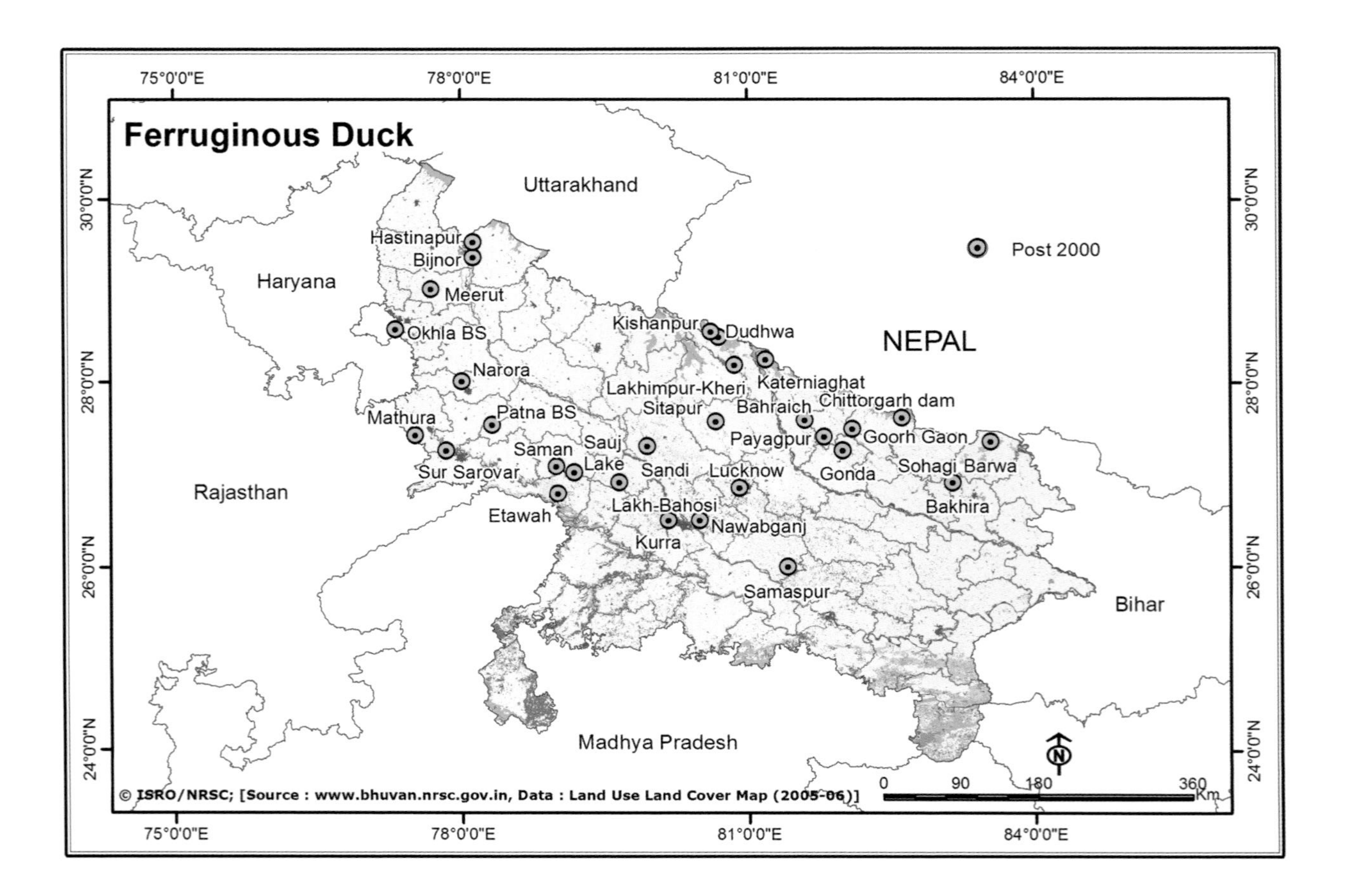

Ferruginous Duck
Uttarakhand
Haryana
Rajasthan
Madhya Pradesh
Bihar
NEPAL
Post 2000
Hastinapur
Bijnor
Meerut
Okhla BS
Narora
Mathura
Patna BS
Sur Sarovar
Saman
Sauj
Lake
Etawah
Kishanpur
Dudhwa
Lakhimpur-Kheri
Kateriaghat
Sitapur
Sandi
Lakh-Bahosi
Kurra
Nawabganj
Lucknow
Bahraich
Payagpur
Chittorgarh dam
Goorh Gaon
Gonda
Sohagi Barwa
Bakhira
Samaspur
0 90 180 360 Km
75°0'0"E
78°0'0"E
81°0'0"E
84°0'0"E
30°0'0"N
28°0'0"N
26°0'0"N
24°0'0"N
© ISRO/NRSC; [Source : www.bhuvan.nrsc.gov.in, Data : Land Use Land Cover Map (2005-06)]

ASAD R. RAHMANI

Sandi is a very well-protected bird sanctuary in Uttar Pradesh where Ferruginous Duck is regularly seen in winter

the following IBAs (Islam & Rahmani 2004): **Lakh-Bahosi** Bird Sanctuary (Farrukhabad), **Narora** (Bulandshahr, Badaun), **Saman** (Mainpuri), **Sandi** (Hardoi), **Bakhira** (Sant Kabir Nagar), **Hastinapur** (Meerut), **Katerniaghat** (Bahraich), **Kudaiyya** (Mainpuri), **Kurra** (Etah, Mainpuri), **National Chambal** WLS (Agra, Etawah), **Nawabganj** (Unnao), **Parvati Aranga** (Gonda), **Payagpur** (Bahraich), **Samaspur** (Raebareli), **Sauj** (Mainpuri), **Sur Sarovar** (Agra), **Surha Taal** (Ballia), **Dudhwa** NP (Lakhimpur-Kheri), **Patna Jheel** (Etah), and **Kishanpur** WLS (Lakhimpur-Kheri).

Gopi Sundar (*pers. comm.* 2006) noted this species in the wetlands of **Etawah** district, and Sharma (1984) based on his three-year study (1980–1983) reported it in **Meerut** district. More recently, Bhargava (2012) recorded it from **Hastinapur** WLS. During surveys between 1995 and 1997, Islam *et al.* (1999) found it in the polishing ponds of the **Mathura Refinery**. Khan (1992) sighted it in the wetlands of **Sitapur** and **Lakhimpur-Kheri** districts. During a three-day survey, Fazlur Rahman (*pers. comm.* 2007) counted 41 Ferruginous Duck in the wetlands of **Kishanpur** Sanctuary. Pasha (1995) reported it from wetlands in **Bijnor** district.

Recent records by Amit Mishra (*pers. comm.* 2010) are as follows: five birds in **Katerniaghat** WLS in January 2008 and four at Raghunath-ka-purva, a small village swamp in **Lucknow** district. Again, 10 birds were sighted in **Sandi** WLS in February 2009 and 15 birds in **Kishanpur** WLS in December 2009. Urfi (2003) recorded about 20 individuals in January 2002 in **Okhla** Bird Sanctuary that lies on the Uttar Pradesh and Delhi border.

During bird surveys at **Narora** in Bulandshahr district, there have been regular sightings from a small wetland called Hakimpur inside the Atomic Power Plant: 20, 09, and 17 birds were seen near Narora Barrage in January 2011, February

2012, and February 2013 respectively (P.D. Mishra & Raja Mandal *pers. comm.* 2010). Regular sightings were reported by Forest staff from **Sandi** WLS during January and February from 2010 to 2013 consecutively, though in small numbers.

Rajat Bhargava and Nikhil Shinde of BNHS during their bird surveys in the Terai found it on the following wetlands: up to 60 on February 7, 2014 in Singhrana Taal, North Chowk, **Sohagi Barwa** WLS; 20–30 on February 2, 2014 in Chittorgarh dam, Rampur Range, **Soheldev** WLS; and 40–50 on February 11, 2014 in Goorhgaon wetland, **Balrampur**.

Ecology: In **India**, it can be seen in shallow ponds, pools, and marshes near vegetated shoreline, large marshes, wetlands, and sometimes in rivers. Prefers shallower and more vegetated areas than other *Aythya* species and seldom sits out in open water. Feeds on seeds, roots, and the green parts of aquatic plants. It also feeds on insects, worms, molluscs, crustaceans, amphibians, and small fish. It often feeds at night, and will upend (dabble) for food, as well as the more characteristic diving.

In India, it was found breeding in some wetlands of **Kashmir** such as **Hokarsar**, **Anchar,** and **Haigam** (Bates & Lowther 1952), but we do not have recent confirmed records of breeding. Earlier it used to breed in such vast numbers that collection of the eggs of this duck and of the Mallard, and bringing them into Srinagar by boat for sale, was a regular and profitable profession for people living in the vicinity of their breeding haunts (Baker 1921).

According to del Hoyo *et al.* (1992), it starts breeding by April and May in Central Europe, singly or in loose groups, making its nest with leaves, stems, and grass on grounds with thick vegetation or in reed beds. It lays 8–10 eggs, incubating them for 25–27 days, and the chicks become sexually mature in one year.

It is a gregarious species, forming large flocks in winter, often mixed with other diving ducks.

Threats: In **India**, it is mainly threatened by habitat destruction and modification, and by trapping and poisoning. In Patna (Bihar), about 10 birds (mainly males) were recorded on sale in 2013 and were said to have come from Unnao via Varanasi, but this is unlikely as Patna gets many waterbirds caught locally (Abrar Ahmed *pers. comm.* 2013). Wherever the birds were caught, it proves that bird trapping, although much reduced, is still a threat to this and other species.

Conservation measures underway: It is protected under the Indian Wildlife (Protection) Act, 1972, which bans its hunting, trapping, trading, and poaching. It is listed in Schedule IV of the Act.

RECOMMENDATIONS

(1) Strict prevention of trapping and poisoning of waterfowl which unfortunately is quite common in some areas.
(2) Development of National Wetland Conservation Act for protection of wetlands.
(3) Better protection of non-protected wetland IBAs.
(4) Regular monitoring of all waterfowl, particularly threatened species.

Lesser Flamingo

Phoeniconaias minor (Geoffroy Saint-Hilaire 1798)

V. GOPI NAIDU

IUCN and BirdLife International (2013) list the Lesser Flamingo as Near Threatened because its populations appear to be undergoing a moderately rapid reduction. It is a marginal bird in Uttar Pradesh with very few records.

Field Characters: The Lesser Flamingo is a large but dainty bird, which stands 90 to 105 cm tall, and has long legs and neck, and deep rose pink plumage. Adult has black-tipped dark red bill, dark red iris and facial skin, and deep rose pink on head, neck, and body. Blood red centres to lesser and median upperwing-coverts are also characteristic. Young ones are overall brown, with grey-brown bill and grey legs. The immature birds have a greyish brown, pale pink body (as they grow, the pinkness increases). Bill coloration also develops with age.

Distribution: In India, the Lesser Flamingo is found regularly in Gujarat, Maharashtra, Rajasthan, and in the western **Uttar Pradesh**-Delhi-Haryana belt, with erratic records from many states (Orissa, Andhra Pradesh, Tamil Nadu). It is found more often in salt lakes and brackish water than the Greater Flamingo *Phoenicopterus roseus*, hence it is more restricted in distribution. Earlier it was mainly reported from north and northwest India, but now it is being increasingly seen in southern India also. Rahmani (2012) has recently collated all important records from India. Although found in very small numbers in some wetlands, we have confirmed records in Uttar Pradesh only from **Patna** Bird Sanctuary.

The Lesser Flamingo is a marginal species in Uttar Pradesh with confirmed records only from Patna Bird Sanctuary

Ecology: The Lesser Flamingo is known to congregate in tens of thousands, often with Greater Flamingo, but sometimes in single-species flocks as the habitat requirements of the two species are slightly different. The Lesser Flamingo prefers heavily saturated brine where it feeds through its specialised bill on microscopic blue-green algae *Spirulina* spp., *Oscillatoria* spp., and *Lyngbya* spp. and diatoms *Navicula* spp., Bacillariophyceae. To a lesser extent, the species also takes small aquatic invertebrates such as rotifers *Brachiomus* spp. (del Hoyo *et al.* 1992) and occasionally insect larvae (Ali & Ripley 1987). It regularly drinks fresh water.

As it is a marginal species in Uttar Pradesh, we are not describing its ecology in detail.

Threats: Hunting is not a major threat in **India** as the bird is not good to eat. However, habitat disturbance and habitat modification are major threats as it is a very specialised feeder. Any change in the water chemistry impacts this bird through its food chain. Therefore, disturbance of habitat through excessive extraction of salt (e.g., **Sambhar Lake**) and pollution (e.g., **Sewri** mudflats, **Mumbai**) greatly disturb this species.

Lesser Flamingo are highly prized in aviculture, and at end level, a pair can fetch Rs. 20,000 to 30,000 in Indian markets. However this practice is becoming less common with more awareness and enforcement (Ahmed 2012).

Conservation measures underway: It is listed in Schedule IV of the Indian Wildlife (Protection) Act, 1972, and also in CITES Appendix II and CMS Appendix II. It is protected in most of the countries where it occurs. It is found in many PAs/IBAs in India. There is an IUCN-SSC/Wetland International Flamingo Specialist Group which publishes research and conservation action plans for all flamingos of the world.

RECOMMENDATIONS

Rahmani (2012) has given recommendations for India. As it is a marginal species in Uttar Pradesh, we are not giving state-specific recommendations. General recommendations are as follows:

(1) Study its movement through satellite tracking.

(2) Monitor the breeding populations.

(3) Toxicological and disease aspects need to be addressed as industrial effluents, pesticides, sewage, and agrochemicals are drained into flamingo habitats.

Painted Stork

Mycteria leucocephala (Pennant 1769)

SANJAY KUMAR

Although it is one of the most abundant Asian storks, and particularly common in the wetlands of Uttar Pradesh, the Painted Stork has been classified as Near Threatened because it is thought to be undergoing a moderately rapid population decline in Southeast Asia, owing to hunting, drainage, and pollution in its habitat (BirdLife International 2013).

Field Characters: A long-legged, long-necked, lanky bird, slightly less than a metre (93 cm) in height, the Painted Stork inhabits marshes and lakes. It has a long, heavy, yellow bill, slightly decurved at the tip, and an unfeathered waxy yellow face. Head, neck, breast, and back are white, with closely barred belly band, and black-and-white wing-coverts. Inner secondaries rose pink, falling over the black tail, give it the name Painted. Legs and feet fleshy, sometimes nearly red, often appearing white due to their habit of defecating on their legs (urohydrosis) especially when at the nest. Sexes alike, but female appears to be smaller. The downy young are mainly whitish with grey bill and blackish facial skin. The juveniles assume a brownish plumage and like most other storks reach breeding condition after two to three years.

Distribution: In India, the Painted Stork is found throughout the plains, rarely in the **Brahmaputra** valley and is not recorded in Andaman and Nicobar Is. It is becoming much more common in south India where many nesting colonies are

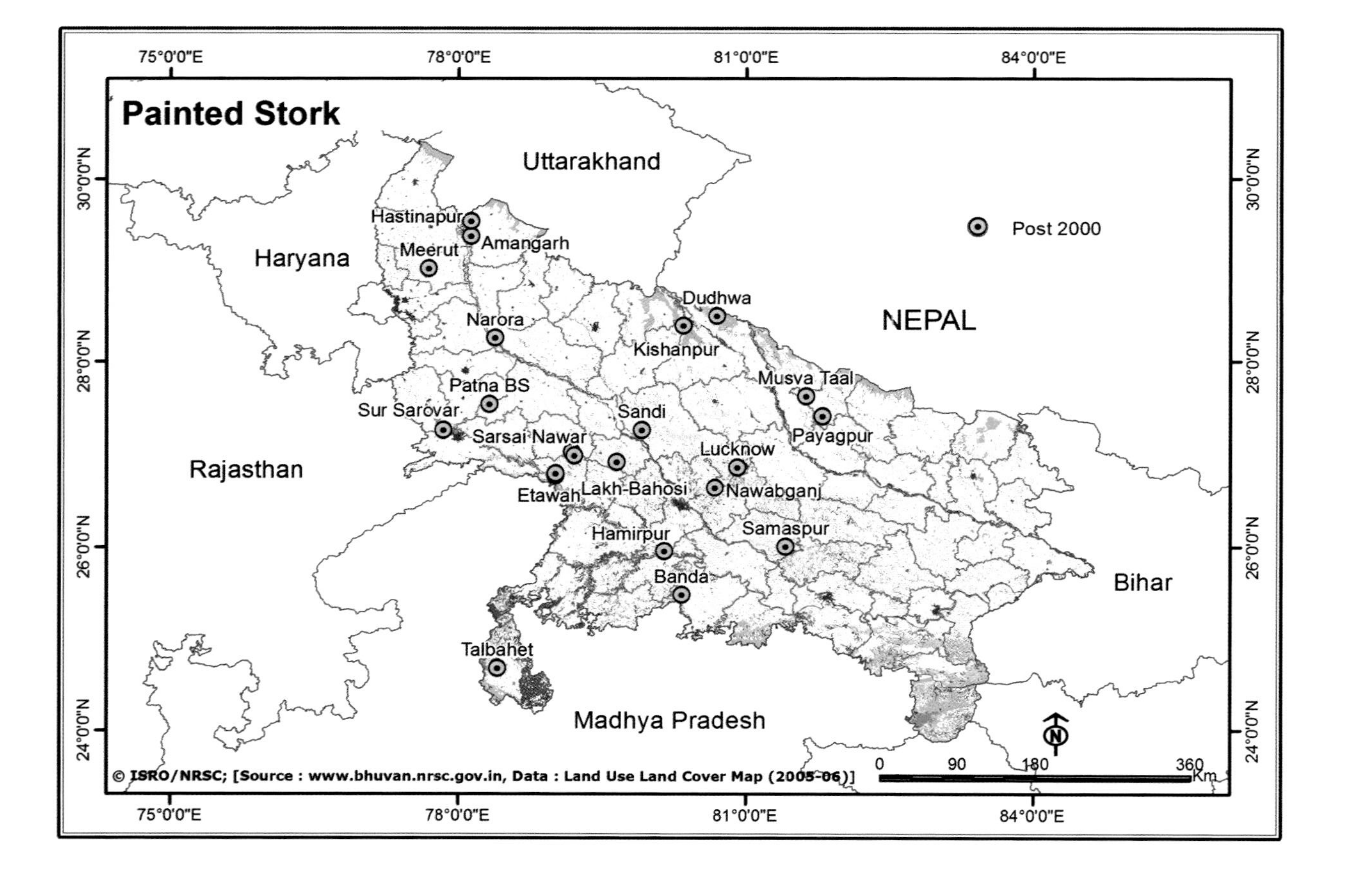
Painted Stork
Post 2000
Uttarakhand
Haryana
NEPAL
Rajasthan
Bihar
Madhya Pradesh
Hastinapur
Amangarh
Meerut
Narora
Dudhwa
Kishanpur
Patna BS
Sur Sarovar
Sandi
Musva Taal
Payagpur
Sarsai Nawar
Lucknow
Etawah
Lakh-Bahosi
Nawabganj
Hamirpur
Samaspur
Banda
Talbahet
0 90 180 360 Km
75°0'0"E
78°0'0"E
81°0'0"E
84°0'0"E
30°0'0"N
28°0'0"N
26°0'0"N
24°0'0"N
© ISRO/NRSC; [Source : www.bhuvan.nrsc.gov.in, Data : Land Use Land Cover Map (2005-06)]

protected by villagers, and it is also seen in sanctuaries. Recently, Rahmani (2012) has collated all the important records from India.

Records from Uttar Pradesh: In **Uttar Pradesh**, it can be seen in any wetland, sometimes even in roadside ditches. However, it has been specifically reported from the following IBAs/PAs: **Kudaiyya marshland** (Mainpuri), **Lakh-Bahosi** Bird Sanctuary (Farrukhabad), **Narora** (Bulandshahr and Badaun), **Nawabganj** Bird Sanctuary (Unnao), **Patna** Bird Sanctuary (Etah), **Payagpur** Jheel/**Bagheltal** (Bahraich), **Samaspur** Bird Sanctuary (Raebareli), **Sauj** Lake (Mainpuri), **Sarsai Nawar** Lake (Etawah), **Sheikha Jheel** (Aligarh), **Sur Sarovar** Bird Sanctuary (Agra). Small breeding colonies are present in many villages. We need to survey all these colonies and estimate how many pairs of Painted Stork breed in Uttar Pradesh. It is found all over the plains in the state.

There are numerous breeding records across the state except the dense forested areas of the Terai, which is not a favoured habitat of this bird. However, there is one sighting by Amit Mishra (*pers. comm*. 2013), when he saw 40 Painted Storks on the waterlogged side of Jhadi Taal in **Kishanpur** WLS. Reports of Painted Stork being seen in large numbers are as follows: >40 birds seen in **Sandi** WLS in February 2008; 65 birds in **Hakimpur** wetlands, **Narora** in February 2008; >150 birds in **Sandi** WLS in May 2009; >50 birds in an agricultural field on Etawah-Mainpuri road in March 2010; >100 birds on the edges of **Pili** reservoir in **Amangarh** Reserve, **Bijnor** district; and in February 2012, >100 birds were spotted in fallow land adjoining **Patna** WLS.

Rajat Bhargava and Nikhil Shende of BNHS during their bird surveys of the Terai found two birds on February 11, 2014 in Musva Taal, **Bahraich**.

In western Uttar Pradesh it is not as common as in eastern UP, nevertheless it is found in **Hastinapur** WLS and Sobhapur area in **Meerut** (Bhargava 2012), and in **Bijnor Barrage** and **Okhla** Bird Sanctuary.

This is one of the few species capable of nesting successfully even in busy urban areas. Amit Mishra recorded such a case in Lucknow, where at Park Road, one of the busiest roads in the vicinity of the Lucknow Zoo, they congregate and nest on some False Ashoka or Mast trees *Polyalthia longifolia* on private land. The counts revealed 70 birds in October 2007; *c.* 100 in February 2008; 50 and 75 in October and December 2008 respectively. When some of the trees were removed to build concrete structures, the nesting population was reduced drastically. For instance, in December 2012, only 40 birds were observed nesting on the remaining trees (Amit Mishra *pers. comm.* 2013).

Ecology: The ecology of the Painted Stork has been studied by many workers including Desai *et al.* (1977), Hancock *et al*. (1992), Ishtiaq (1998), Urfi (1996, 2003), Urfi & Kalam (2006), Urfi *et al*. (2007), Kalam & Urfi (2008), and Meganathan & Urfi (2009). It frequents freshwater marshes, lakes and reservoirs, flooded fields, rice paddies, freshwater swamp forest, river banks, intertidal mudflats, and salt pans. It forages in flocks in shallow waters along rivers or lakes. It immerses its half-open beak in water and sweeps it from side to side, snapping up its prey of small fish that are sensed by touch. As it wades along it also stirs the water with its feet to flush hidden fish. It nests colonially in trees, often with other waterbirds. The only sounds it produces are a weak moan or by bill-clattering. It makes short-distance movements in some parts of its range in response to food and for breeding. Like other storks, it is often seen soaring on thermals.

It breeds colonially in single-species or mixed heronries and if not molested, such heronries become traditional. The birds arrive just before the monsoon breaks and spend considerable time on the selected nesting trees, perhaps waiting for the right cue to start making nests of sticks and leaves. Mating frequently takes place on the nest or a nearby branch, and the female lays three or four, rarely five eggs. Both parents incubate and rear the chick.

Threats: The increasing impact of habitat loss, disturbance, pollution, drainage, and hunting of adults and collection of eggs and nestlings from colonies is a cause for concern in many range countries. In India, though it is protected traditionally in many areas, poaching by tribal and amateur hunters and pesticide poisoning are major threats. Nest predation by mammalian and avian predators is the major threat in some colonies, aggravated by human interference (Sundar 2005b). During winter, it is illegally caught using leg nooses in certain villages around Kanpur, Lucknow, and Raibareli, and sold for food and occasionally for aviculture or zoos (Ahmed 2012).

Conservation measures underway: It is listed in Schedule IV of the Indian Wildlife (Protection) Act, 1972. Further, its nesting sites are traditionally protected, as a result of which its population is increasing in some areas. It also occurs in a number of PAs/IBAs.

RECOMMENDATIONS

Rahmani (2012) has given general recommendations for India, many of which are valid for Uttar Pradesh.

(1) Conduct studies on the level of threat to this bird due to pesticide poisoning of its food.

(2) Conduct state-wide surveys every two years to monitor its populations, particularly at major nesting colonies in Uttar Pradesh.

(3) To study movement and dispersal, conduct satellite telemetry studies in major breeding areas.

(4) Bring about strong legislation to protect all types of wetlands.

Black-necked Stork

Ephippiorhynchus asiaticus (Latham 1790)

K.S. GOPI SUNDAR

Based on information gathered by BirdLife International (2013), IUCN includes Black-necked Stork in the Near Threatened category as this species has undergone a moderately rapid overall population reduction, which is projected to continue. The population estimate varies between 10,000 and 20,000 in the whole world.

Field Characters: A characteristically large bird between 130 and 150 cm tall, with bright red legs, white body, extensive black on the wings and tail, and notably an iridescent black head and neck with large black bill. The young looks like a washed-out version of the parents, with dull brown replacing glossy black parts, and dirty white replacing pure white. Genus *Ephippiorhynchus* is unique among storks in exhibiting sexual dimorphism in coloration: iris dark brown in male and yellow in female. Like most storks, the Black-necked Stork flies with its neck outstretched, not retracted like a heron. The wingspan measures up to 230 cm.

Distribution: The Black-necked Stork is found in South and Southeast Asia, but no where is it common. In India, it is found all over the Indian plains in wetlands, shallow river beds, and mangrove swamps. Recently, Rahmani (2012)

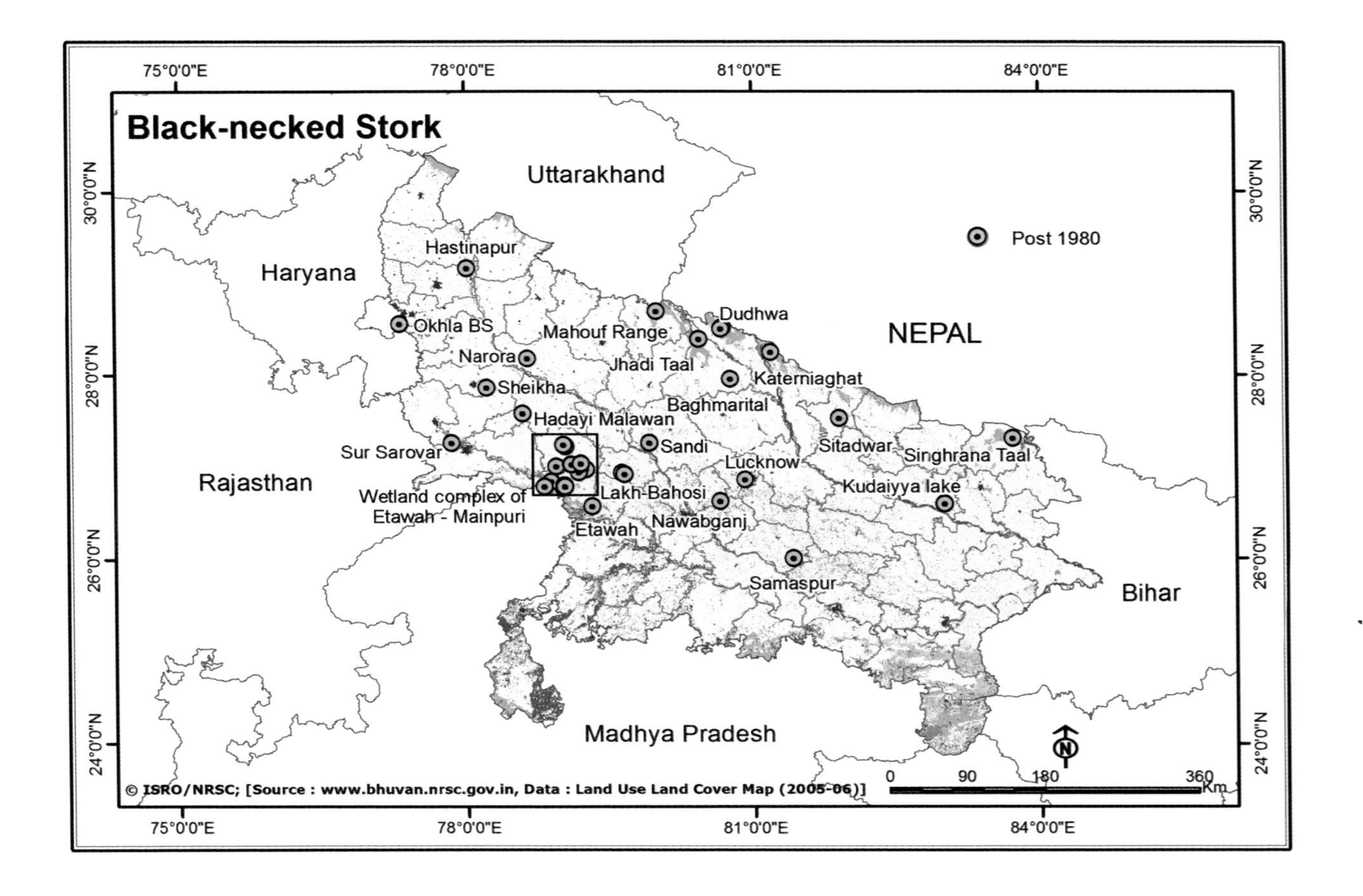
Black-necked Stork
Uttarakhand
Haryana
Rajasthan
Madhya Pradesh
NEPAL
Bihar
Post 1980
Hastinapur
Okhla BS
Narora
Sheikha
Mahouf Range
Jhadi Taal
Dudhwa
Katerniaghat
Baghmarital
Hadayi Malawan
Sandi
Sur Sarovar
Wetland complex of
Etawah - Mainpuri
Etawah
Lakh-Bahosi
Nawabganj
Lucknow
Sitadwar
Singhrana Taal
Kudaiyya lake
Samaspur
75°0'0"E
78°0'0"E
81°0'0"E
84°0'0"E
30°0'0"N
28°0'0"N
26°0'0"N
24°0'0"N
0
90
180
360
Km
© ISRO/NRSC; [Source : www.bhuvan.nrsc.gov.in, Data : Land Use Land Cover Map (2005-06)]

ASAD R. RAHMANI

Black-necked Stork regularly breeds in Dudhwa National Park and some wetland reserves of Uttar Pradesh

has given important site records from the Indian subcontinent. Here we give records from **Uttar Pradesh**.

Records from Uttar Pradesh: Perhaps the best known population is present in **Etawah** and **Mainpuri** districts. Based on his surveys and road transects from December 2000 to February 2002, Sundar (2005a) estimated a population of 200–250 individuals, i.e., at least 20% of the total estimated Indian population. In the mid 1990s, 19 were seen in **Dudhwa** NP (Maheswaran 1998). During a 2010 survey of Gharial in the River Ganga in **Hastinapur** WLS, along a 195 km stretch from Bijnor Barrage to Narora Barrage, 30 Black-necked Stork were counted (Sanjeev Kumar Yadav *pers. comm*. 2010). These birds, which are often seen in pairs, were easily sighted during a full day bird survey by boat conducted through the Meerut Forest Department from **Hastinapur** WLS towards Jyotiba Phule Nagar in 2012, totalling *c*. 15 birds (Bhargava 2012, Lalit Kumar Verma *pers. comm*. 2012).

Dudhwa, **Kishanpur**, **Katerniaghat**, **Suhelwa,** and **National Chambal WLS** are other important IBAs for this species. During summer, when most wetlands are dry, up to 15 Black-necked Stork can be sighted occasionally in Jhadi Taal of **Kishanpur** WLS. In June 2009, 12 Black-necked Stork (with 2 juveniles) were sighted on the dry riverbed of Suheli river in **Dudhwa** NP. Another 12 birds were sighted in May at Jhadi Taal in **Kishanpur** Wildlife Sanctuary (Jaswant Singh Kaler *pers. comm.* 2013). Such congregations are not common, except during summer when smaller wetlands are dry. In some protected areas, this bird was seen sharing its habitat with Sarus Crane, but as it is not as common and abundant as

Sarus, sighting outside protected areas is rare. Regular nesting is recorded from **Sandi**, **Samaspur**, **Patna**, and **Sheikha** wetlands. Sporadic records are available from other parts of the state. Black-necked Stork was reported from Hakimpur wetland in the exclusion zone of **Narora** Atomic Power Plant, Bulandshahr, the maximum number being seven in 2011 and 2012. Nesting on a *Ficus* tree on the banks of River Ganga, near Narora Atomic Power Plant, was also reported (P.D. Mishra & Raja Mandal *pers. comm.* 2013).

The Black-necked Stork has been reported from the following IBAs/PAs: **Kudaiyya** Marshland (Mainpuri), **Lakh-Bahosi** (Farrukhabad), **Narora** (Bulandshahr, Badaun), **Nawabganj** (Unnao), **Patna Jheel** (Etah), **Payagpur Jheel** (Bahraich), **Saman** WLS (Mainpuri), **Samaspur** Bird Sanctuary (Raebareli), **Sandi** Bird Sanctuary (Hardoi), **Sauj** Lake (Mainpuri), **Sarsai Nawar** Lake (Etawah), **Sheikha Jheel** (Aligarh), **Sohagi-Barwa** WLS (Gorakhpur), **Sur Sarovar** Bird Sanctuary (Agra). On February 14, 2014, one was seen in flight in Mahauf Range, **Pilibhit** RF by Nikhil Shende of BNHS.

In February 2014, Sanjay Kumar sighted a pair in Pili Dam next to **Amangarh** RF in **Bijnor** district. Another interesting sighting is of 14 Black-necked Storks at one spot in Jhadi Taal in **Kishanpur** WLS in May 2012. In summer, when smaller wetlands dry up, a number of large waterbirds congregate in Jhadi Taal, and sighting of 2–3 pairs and a few juveniles of Black-necked Stork is not unusual but sighting of 14 birds together is a record.

Ecology: Its ecology and breeding biology were studied by Ishtiaq (1998) in Keoladeo NP, by Maheswaran (1998) in **Dudhwa** NP and by Sundar (2003) in

NEERAJ SRIVASTAV

Overfishing with small mesh-size net is a major problem all over Uttar Pradesh, leaving very little food for piscivorous birds such as the Black-necked Stork

Etawah and **Mainpuri** districts. The Black-necked Stork prefers large marshes and jheels, and margins of large rivers and brackish lagoons where it feeds on fish, frogs, snakes, small turtles, injured and unwary birds, and any animal which it can swallow (Ishtiaq *et al.* 2010). Chauhan & Andrews (2006) found that it regularly feeds on eggs of the Three-striped Roof Turtle *Kachuga dhongoka* in the National Chambal Sanctuary. Interestingly, local people appeared to be familiar with the predatory behaviour of this species. Black-necked Stork in Australia is also known to feed on freshly hatched marine turtles at night.

It is generally found in pairs, even outside the breeding season, and pairs maintain large feeding territories throughout the year. It is very rare to see three or four adult birds together. In summer months, when jheels and ponds dry up and food is reduced, it becomes more aggressively territorial and frequently fights over food and space (Maheswaran & Rahmani 2001). However, where food is abundant as in Jhadi Taal, Kishanpur WLS, congregations of 10–14 may be seen. It has the characteristic stork habit of soaring and circling aloft in the heat of the day.

The nest is built on large trees, mostly near water. If left undisturbed, the same tree is used year after year. Ishtiaq *et al.* (2004) found that some nests were reused in **Keoladeo** NP. Pairs spend considerable time on the nest, and mating also takes place there. The female lays two to four eggs, and both parents incubate and raise the chicks. Generally one to three chicks are raised (Sundar 2003), but in good habitats (e.g., **Dudhwa** NP, **Kishanpur** WLS) with plenty of food, it is not uncommon to see four juveniles with parents in some years. Ishtiaq (1998), during her studies in **Keoladeo** NP, found three juveniles raised by a pair in 1996–1997. Similarly, Bhatt (2006) reported a pair with four fledged juveniles in **Marine** NP, **Gujarat** and Sundar *et al.* (2007) found four chicks in **Etawah** and **Mainpuri** with some pairs.

Sundar (2003) studied the breeding biology of this species for three consecutive breeding seasons (1999–2002) in **Etawah** and **Mainpuri** districts. He found 29 pairs in the study period in an area of 500 sq. km. Nests were found even in densely populated areas, frequently close to roads and human habitation. Most pairs raised two chicks, but most of them raised chicks in only one of the three years, while only one pair successfully raised chicks in two consecutive years. The young remained in their natal territories for 14–18 months, but some remained up to 28 months.

Threats: The main threat to the Black-necked Stork is destruction and degradation of its habitat, and also overfishing. Hunting, at least in **India**, is not the main problem, but trapping for zoos was at one time a major problem (Rahmani 1989). Since the ban on bird trade in India in 1991, trapping is somewhat reduced. Sundar (2003) found this species successfully nesting in human-dominated landscape in **Etawah** and **Mainpuri** districts of **Uttar Pradesh**.

The most insidious threat to this grand bird is from conversion of marshlands to agriculture fields, drainage of wetlands for commercial purposes (e.g., aquaculture), agricultural pesticide runoff, and other forms of pollution. Intensive fishing of drying up ponds, their main foraging areas during summer, by very fine mesh nets or zero-net fishing (where even fish eggs and fingerlings are netted), and poisoning, all deplete its food supply. In the face of wetland reclamation, flooded rice paddies have become important, as they may be promoting the dispersal of young birds and preventing fragmentation into sub-populations (Sundar 2003).

Despite the ban on wild bird trade, this species is collected by professional bird traders mainly for zoos, and is also hunted for meat especially in eastern UP and Bihar. According to field surveys conducted by TRAFFIC India on illegal bird trade, it is estimated that on an average, 10 or more Black-necked Stork (especially chicks) appear with particular bird dealers each year in the Lucknow-Kanpur-Allahabad belt, from where they are further retailed to prior orders throughout India. In rare instances, they are said to be smuggled to other countries, especially via the Nepal and Bangladesh borders and further via Pakistan. A pair of Black-necked Stork can sell for Rs. 15,000 to 20,000, while poached birds sell for as low as Rs. 500 to 1500 (Ahmed 1997, Ahmed 2012). Often, fully-fledged chicks are collected and raised for sale as they become quite tame and easy to handle.

Conservation measures underway: It is listed in Schedule I of the Indian Wildlife (Protection) Act, 1972 and also included in CITES Appendix I. In **India,** it occurs in a number of PAs/IBAs. Detailed research has been done in India and is ongoing in Australia.

RECOMMENDATIONS

(1) Systematic surveys should be conducted periodically in the whole state. Perhaps this could be combined with the annual Sarus Survey conducted by the Forest Department.

(2) More attention should be given to the Black-necked Stork populations outside PAs and IBAs.

(3) Strict control on the use of harmful pesticides, particularly near wetlands.

(4) The degree to which habitat fragmentation or loss might affect Black-necked Stork is related, in part, to their movements, about which almost nothing is known. Tracking individuals using satellite telemetry would greatly assist in conservation efforts.

(5) Initiate an active advocacy programme to educate farmers on the importance of wetland birds and their protection.

(6) Total implementation of the ban on hunting and trapping of these birds.

(7) Involve IBCN members in monitoring Black-necked Stork nesting and foraging sites.

(8) Make Black-necked Stork, along with Sarus Crane, an icon of healthy wetlands.

Black-headed Ibis
Threskiornis melanocephalus (Latham 1790)

SANJAY KUMAR

In India, the Black-headed Ibis is widespread and locally common in all the wetter parts of the country, less common in the east. In Uttar Pradesh, it is widespread and very common in marshes and irrigated crop fields. Based on the assessment done by BirdLife International (2013), IUCN has listed the Black-headed Ibis in Near Threatened category as it is undergoing population decline in many countries due to hunting and disturbance at breeding colonies, and drainage of wetlands for agriculture.

Field Characters: A large, domestic hen-sized white bird, with black neck, naked black head, and long stout downcurved black bill. Legs and feet are also black. Adult breeding birds have white lower neck plumes, variable yellow wash to mantle and breast, and grey on scapulars and elongated tertials. Immature birds have white chin and neck, naked face, bare skin around the eye, while the rest of the head and neck are feathered. Sexes alike.

Distribution: The Black-headed Ibis, also called White Ibis, is widespread and extending its range in many parts of India (e.g., Thar Desert) due to the development of canal irrigation. It is resident, nomadic, and local migratory, depending upon the availability of water. Rahmani (2012) has collated recent data from India.

Records from Uttar Pradesh: It is widespread and can be seen in any wetland and irrigated field. Among the IBAs/PAs, it is specifically reported from the following: **Kudaiyya Marshland** (Mainpuri), **Lakh-Bahosi** Bird Sanctuary

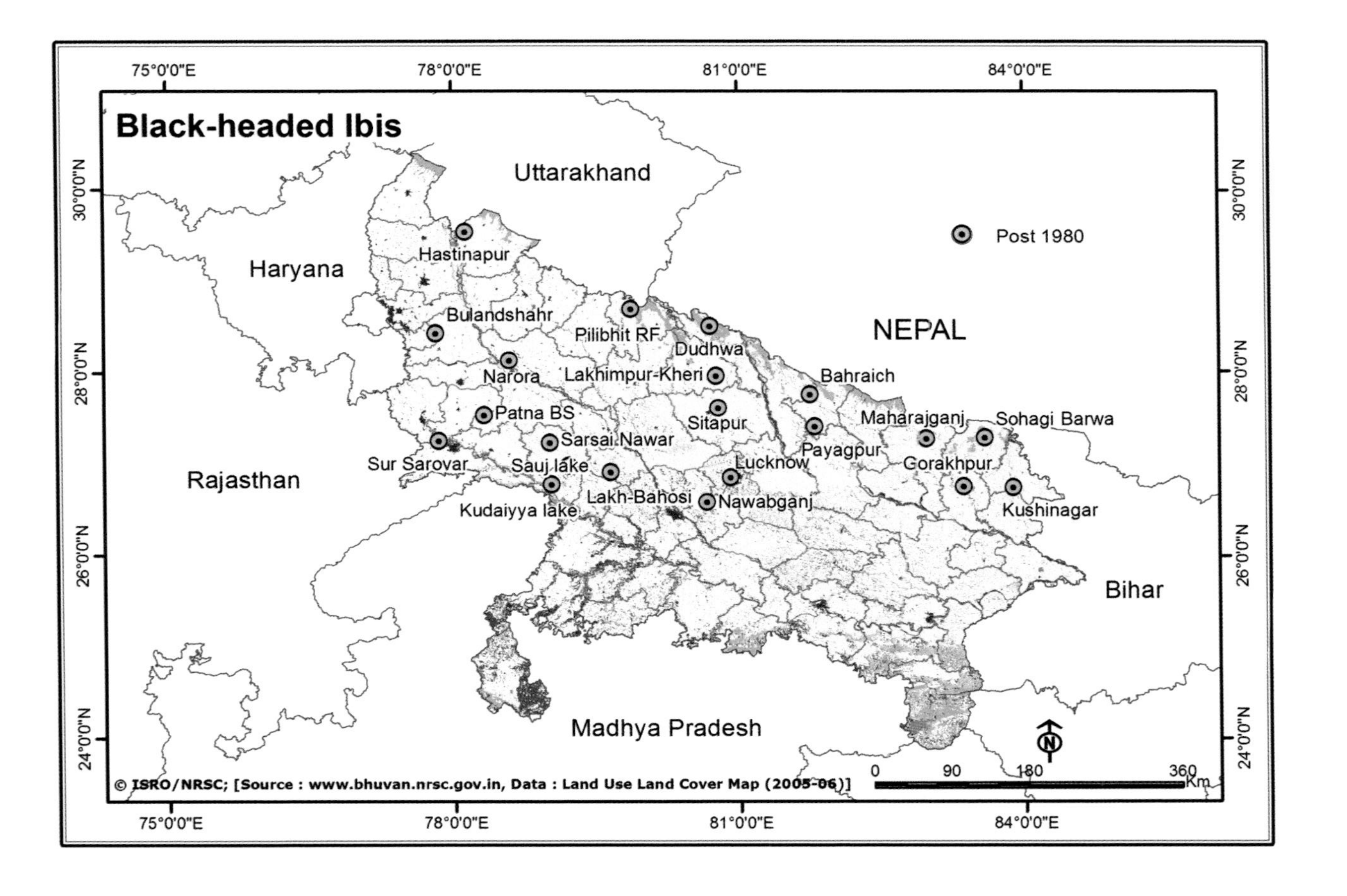
Black-headed Ibis
Uttarakhand
Haryana
Rajasthan
Madhya Pradesh
NEPAL
Bihar
Post 1980
Hastinapur
Bulandshahr
Narora
Patna BS
Sur Sarovar
Sarsai Nawar
Sauj lake
Kudaiyya lake
Lakh-Bahosi
Pilibhit RF
Dudhwa
Lakhimpur-Kheri
Sitapur
Lucknow
Nawabganj
Bahraich
Payagpur
Maharajganj
Gorakhpur
Sohagi Barwa
Kushinagar
75°0'0"E
78°0'0"E
81°0'0"E
84°0'0"E
30°0'0"N
28°0'0"N
26°0'0"N
24°0'0"N
0
90
180
360
Km
© ISRO/NRSC; [Source : www.bhuvan.nrsc.gov.in, Data : Land Use Land Cover Map (2005-06)]

Hundreds of Black-headed Ibis breed in Sur Sarovar Bird Sanctuary near Agra

(Farrukhabad), **Narora** (Bulandshahr, Badaun), **Nawabganj** Bird Sanctuary (Unnao), **Patna Bird** Sanctuary (Etah), **Payagpur/Bagheltal** Jheel (Bahraich), **Saman** WLS (Mainpuri), **Samaspur** Bird Sanctuary (Raebareli), **Sandi** (Hardoi), **Sauj** Lake (Mainpuri), **Sarsai Nawar** Lake (Etawah), **Sheikha Jheel** (Aligarh), **Sohagi Barwa** WLS (Maharajganj), **Sur Sarovar** Bird Sanctuary (Agra) and **Hastinapur** WLS (Meerut). Besides IBAs/PAs, it can also be seen in lowlands/marshlands of districts **Gorakhpur**, **Maharajganj**, **Kushinagar**, **Bahraich**, **Lakhimpur-Kheri**, **Sitapur**, and **Pilibhit**.

This bird is fairly common in the whole state except some areas of Shivalik Hills and Bundelkhand. It can aptly be called a bird of marshland. Wherever a little water or marshy patches are available, sighting this bird is almost certain.

Another characteristic feature of this bird is its high adaptability to changes in the habitat conditions. It is gregarious and known for its ability to live harmoniously in large communities, and many past sightings are of large congregations of birds, especially at the time of breeding. In Uttar Pradesh, this ibis is found in all the protected areas, but for nesting records, no place can surpass **Sur Sarovar** Wildlife Sanctuary in Agra district. From June onwards, one can witness a nesting carnival here. This *sarovar* or waterbody is a centre of intense nesting activity primarily by the Black-headed Ibis, along with other egrets and herons. Almost all the vegetation on the western side of the lake (mostly *Prosopis juliflora*) is occupied for communal nesting. If observed from a distance, the trees appear to be festooned with white dots. More than 500 nests were recorded during the 2010–11 nesting season. It was also noticed that the Black-headed Ibis prefers the top or middle storey of a tree for making its nest.

A second nesting in case of unsuccessful breeding was also reported. This way, the whole process was prolonged till September. Another remarkable nesting site was observed at Nawab Ali-ka-purva, a hamlet some 24 km from **Lucknow** on the Lucknow-Sultanpur highway, where, under the care and protection of locals, these birds gather in large numbers to nest. Nesting was observed here till November, up to 200 birds being recorded in November 2008 (Amit Mishra *pers. comm.* 2010). Despite the site being right alongside a highway, with all sorts of sound pollution, the birds breed successfully. After the breeding season is over, they segregate and form comparatively small groups.

In **Sitarasoi** wetland in **Sitapur** district, a group of 50 birds was recorded throughout the month of March as the waterbody almost dried up with mere traces of water remaining (Kumar & Srivastava 2011). Between 35 and 50 birds were regularly reported from **Hakimpura** wetlands within the exclusion zone of Narora Atomic Power Plant, in **Bulandshahr** district, nesting and perching on an *Acacia* tree alongside the canal (P.D. Mishra & Raja Mandal *pers. comm.* 2013).

Rajat Bhargava and Nikhil Shinde of BNHS during their bird surveys of the Terai found five birds in Chittorgarh dam, Rampur Range, **Soheldev** WLS on December 12, 2013.

Ecology: The Black-headed Ibis is found in all sorts of wet areas, from paddy fields, freshwater marshes, lakes, rivers, flooded grasslands, to tidal creeks, mudflats, brackish marshes, and coastal lagoons, from lowlands to 950 m. It is gregarious, mixing easily with other waders such as storks, egrets, spoonbill, and small waders. It is never found far from water. It feeds almost entirely on animal matter, fish, frogs, aquatic insects, crustaceans, and worms, the last two generally probed out from squelchy mud by it downcurved bill.

It nests colonially with other heronry species, during the monsoon in partially submerged thorny trees to avoid ground predators. A platform nest is made where two to four eggs are laid. Incubation and chick rearing are shared by both parents. Heavy predation of eggs and small chicks by House Crow *Corvus splendens* and pre-fledged chicks by eagles has been noted. Where unmolested, it nests on trees growing even in crowded villages, sometimes away from water, with other colonial nesters such as Painted Stork, Grey Pelican, and egrets.

Threats: It suffers from the usual threats common to all wild birds dependent on natural wetlands in South and Southeast Asia: drainage, disturbance, pollution, conversion of habitats to agriculture, hunting, and collection of eggs and nestlings from colonies. A combination of these factors has probably caused its decline in some countries.

This species is mainly trapped and hunted for meat and occasionally sold for the zoo trade and private aviculture, especially in the eastern UP belt and parts of Bihar. During the breeding season, the adult birds and full-fledged chicks are poached by travelling nomadic tribes such as Gulgulawas and Kurmi-Baheliyas, mainly for food. They are also caught using leg nooses placed near small wetlands

Intensive fishing using nylon nets with very small mesh-size (mosquito nets) depletes all types of aquatic food of Black-headed Ibis

by professional bird trappers such as Mirshikaris and Baheliyas. In some places in Uttar Pradesh, such as Raebareli and Unnao, a stuffed Black-headed Ibis or a Woolly-necked Stork is placed near the nooses to lure passing wild ibises into the trapping range. The Black-headed Ibis, being resident, is trapped for waterbird trade throughout the year in eastern UP wherever there are settlements of Pathami trappers, some of whom are expert in trapping waterbirds (Ahmed 1997, 2012).

Conservation measures underway: It is listed in Schedule IV of the Indian Wildlife (Protection) Act, 1972 so its hunting in India is totally prohibited. It is found in many PAs/IBAs and also gets protection due to religious and traditional practices in many areas (e.g., temple tanks).

RECOMMENDATIONS

(1) Strict enforcement at the grassroots level to prevent trapping and poaching of this species, particularly during the breeding season.

(2) Conduct survey involving members of BNHS, IBCN, other conservation organizations, and civil society to identify and protect important breeding colonies. Subsequently, conduct surveys of identified breeding colonies every two years to monitor the population trends.

(3) Study the impact of pesticides on its food chain.

(4) Study the impact of mobile phone towers on the breeding colonies (many colonies are present inside towns).

(5) Conduct ringing, colour banding, and satellite tracking studies to determine its movements.

Spot-billed Pelican

Pelecanus philippensis Gmelin 1789

VINAYAK YARDI

In 2001, BirdLife International listed the Spot-billed Pelican as Vulnerable as it had declined at a moderately rapid rate owing to a number of threats. However, increased protection, mainly in India, has since enabled recovery in its numbers, and the status of the species was downlisted from Vulnerable to Near Threatened in 2007.

Field Characters: The Spot-billed Pelican, though it is the smallest pelican in India, is still a large bird (length *c.* 140–152 cm). It is mainly drab white, with spotted bill and pouch, and dusky, curly feathers on hind crown and hind neck. The large bill is pinkish with spots, which appear only after a year. The full adult breeding plumage appears in the third year. In flight, primaries are dusky and secondaries dark from below. It is easily confused with the two other pelicans found in India: Dalmatian Pelican *P. crispus* is larger, and a brighter white with orange pouch and bushy, curly crest; the juvenile Great White Pelican *P. onocrotalus* is larger with darker head, neck and upperparts, paler lores and blackish flight feathers. At a distance, it is difficult to differentiate from other pelicans in the region although it is smaller, but at close range the spots on the upper mandible, the lack of bright colours, and the greyer plumage are distinctive.

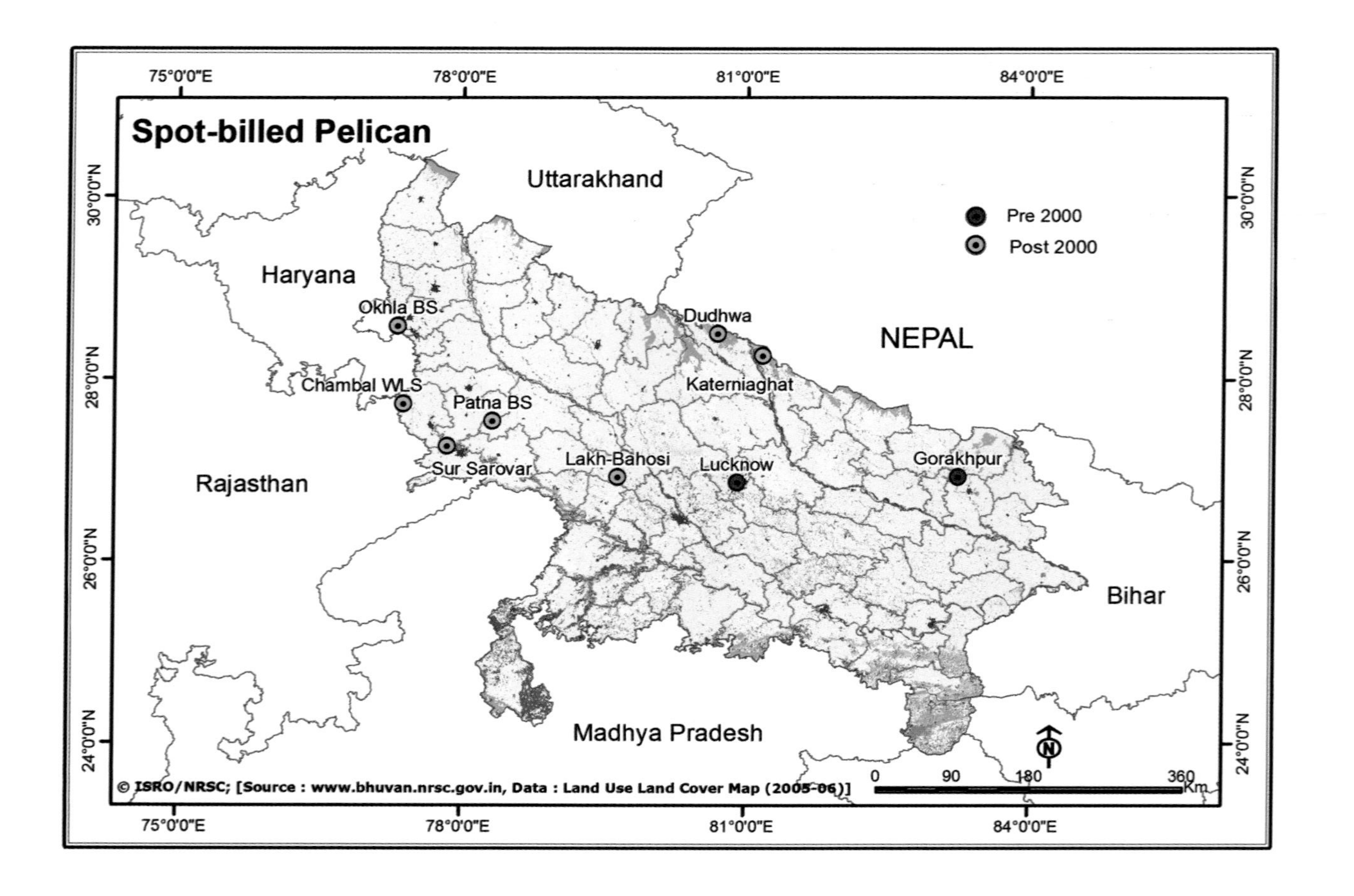

Spot-billed Pelican
Pre 2000
Post 2000
Uttarakhand
Haryana
NEPAL
Rajasthan
Madhya Pradesh
Bihar
Okhla BS
Chambal WLS
Patna BS
Sur Sarovar
Lakh-Bahosi
Lucknow
Dudhwa
Katerniaghat
Gorakhpur
0 90 180 360 Km
75°0'0"E
78°0'0"E
81°0'0"E
84°0'0"E
30°0'0"N
28°0'0"N
26°0'0"N
24°0'0"N
© ISRO/NRSC; [Source : www.bhuvan.nrsc.gov.in, Data : Land Use Land Cover Map (2005-06)]

Distribution: The Spot-billed Pelican is a resident and local migrant in well-watered tracts of South and Southeast Asia. In **India**, it mainly breeds in the south (Subramanya 1996a,b) and the Brahmaputra valley (Talukdar 1995, 1999), but can be seen in the non-breeding season in large wetlands in other parts. In **Uttar Pradesh**, it is seen in the following large wetlands: **Sur Sarovar (Agra)**, **Patna Jheel (Agra)**, **Lakh-Bahosi (Farrukhabad)**, **Okhla** Bird Sanctuary **(Gautam Buddh Nagar)**, **National Chambal** Sanctuary **(Etawah)**, and occasionally in **Dudhwa** and **Katerniaghat (Lakhimpur-Kheri)**. It is also seen on the River **Yamuna** near Sur Sarovar Bird Sanctuary, and may be present on many larger rivers and large irrigation wetlands.

Ecology: The Spot-billed Pelican inhabits a variety of deep and shallow wetlands, both man-made and natural, freshwater and saline, open and forested. It forages and flies in small flocks and breeds colonially in tall trees. The nests are usually built alongside those of other colonial waterbirds, particularly egrets, cormorants, and Painted Stork. Its main food is fish, but it also catches frogs, aquatic snakes, and insects. Some populations appear to be sedentary, but they need to be studied with banding and telemetry.

The Spot-billed Pelican nests on small to moderately tall trees, generally submerged in water, in some places even on Palmyra palms *Borassus flabellifer*. It builds a platform nest and both parents help in nest building and protection. Female lays two to five eggs, and both parents incubate and help in chick rearing. The nestling has pure white down, exchanged for greyish speckled plumage as it grows. The spots on the bill appear only after a year. The full adult breeding plumage appears in the third year. Its breeding behaviour and ecology have been studied in great detail (Neelakantan 1949, Nagulu 1983, Manakadan & Kannan 2003).

Threats: In India, hunting is not the main threat as the species is protected by law and also by local sentiment. However, its wetland habitat is under tremendous human pressure, particularly from the fishing community. Fishing is intense in almost all wetlands, sometimes with very narrow mesh-size (zero) nets which catch even fish eggs and fingerlings. Drainage of natural wetlands for agriculture is another problem, which will be further aggravated by climate change in future. Commercial aquaculture in natural and man-made wetlands and competition between birds and fishermen result in some persecution and disturbance. Although its numbers may have increased in the last decade, possibly due to a combination of more intensive surveys/counts and better protection to nesting colonies, Manakadan & Kannan (2003) foresee population declines in future due to the multitude of anthropogenic pressures on pelicans, especially at their foraging grounds.

During winter, this species is trapped and hunted (in low numbers) in north and northeast India for meat. The Gulgulawa tribe of Jharkhand and Kurmi-Baheliyas poach the chicks in northern India for their own consumption and some local sale. In eastern **Uttar Pradesh** and Bihar, it is occasionally captured with plastic nooses and sold for meat and occasionally for zoos. At the local village level, a pelican can fetch Rs. 1500 to 2000. In south India, the Narikurava tribals hunt and poach the birds occasionally (Bhargava 2012).

Conservation measures underway: The Spot-billed Pelican is protected by Schedule IV of the Indian Wildlife (Protection) Act, 1972, and its hunting and shooting is strictly prohibited. Even disturbance to its nesting sites is prohibited under the law. Although it nests and forages in some PAs/IBAs, most of the nesting colonies are protected by local community initiatives (Manu & Jolly 2000).

RECOMMENDATIONS

Rahmani (2012) has given recommendations for the whole country, many of which are valid for Uttar Pradesh:

(1) Conduct an all-India survey of its nesting colonies every three years to monitor the population trend.

(2) Protect wetlands in and around nesting colonies.

(3) Intensive patrolling of sites known to have poaching pressure; establishment of a rehabilitation programme for traditional hunting communities.

(4) Ban on fishing practices and gear that are non-sustainable and destructive to fish populations.

(5) Establishment of artificial foraging sites to alleviate conflict with fishermen. These could even be in the form of simple fish farms established by the Forest Department to cater to the food requirement of breeding colonies.

(6) Where power lines are a problem, the Forest Department needs to look into the technicalities of stopping or minimising deaths by electrocution.

(7) Tourists visiting pelicanries should be regulated to ensure minimal disturbance, and steps should be taken to reduce other negative impacts of tourism (e.g., litter).

Oriental Darter

Anhinga melanogaster Pennant 1769

VINAYAK YARDI

Based on the assessment done by BirdLife International (2013), the IUCN places this species in the Near Threatened category because of a moderately rapid decline in its population in some countries owing to pollution, drainage, hunting, and collection of eggs and nestlings. It may not be so rare in Uttar Pradesh, but its global population is declining, particularly in Southeast Asia.

Field Characters: A sleek waterbird, mainly black in adult stage, like the cormorants but with longer, more slender snake-like neck, narrow head and long, straight, pointed bill. Tail long, stiff, and fan-shaped when spread. Head and neck velvety chocolate-brown with white chin and throat and a narrow white line from below eye halfway down each side of neck. Back and wings have longitudinal silvery grey streaks. Legs and webbed feet are black, lower mandible flesh-coloured. Sexes alike. Young ones have white down which persists on some parts till almost fledged. Immature birds are dark brown above, with paler neck and head, and below dark brown, almost black. Leg and feet pale. Mantle also dull, streaked with rufous and silvery grey.

Distribution: The Oriental Darter or Snakebird is widespread in suitable wetlands in South and Southeast Asia. In **India** it is found from coastal wetlands to about *c*. 300 m in the Himalaya. Its only requirement is clear, deep, unpolluted water with plenty of fish. It is also found in jheels with deep pools of 1–2 m, larger rivers, and man-made reservoirs. Rahmani (2012) has recently given important sight records of India.

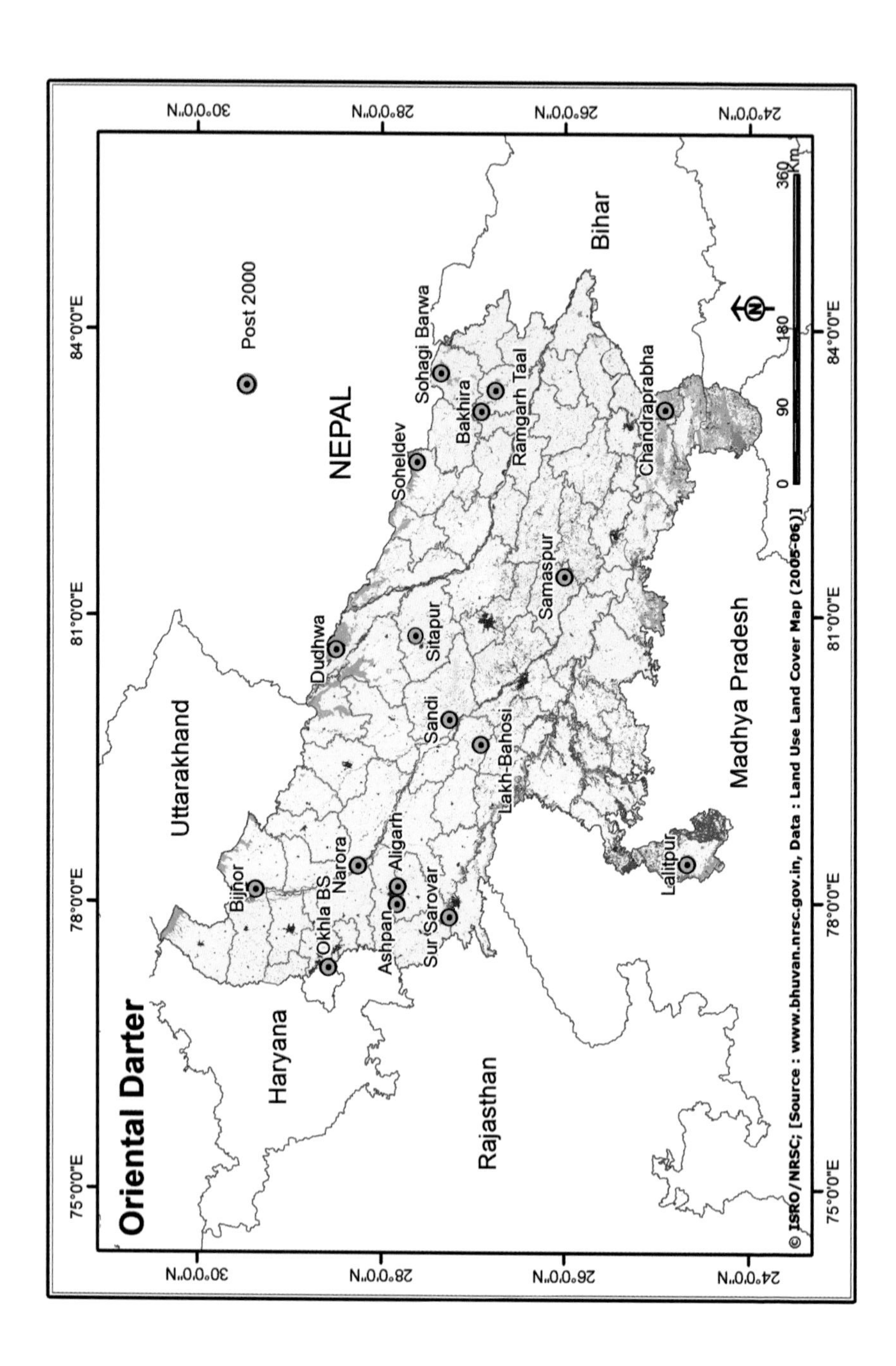
Oriental Darter
Post 2000
NEPAL
Uttarakhand
Haryana
Rajasthan
Madhya Pradesh
Bihar
Bijnor
Okhla BS
Narora
Aligarh
Ashpan
Sur Sarovar
Lakh-Bahosi
Sandi
Dudhwa
Sitapur
Samaspur
Soheldev
Sohagi Barwa
Bakhira
Ramgarh Taal
Chandraprabha
Lalitpur
75°0'0"E
78°0'0"E
81°0'0"E
84°0'0"E
30°0'0"N
28°0'0"N
26°0'0"N
24°0'0"N
0
90
180
360
Km
© ISRO/NRSC; [Source : www.bhuvan.nrsc.gov.in, Data : Land Use Land Cover Map (2005-06)]

Records from Uttar Pradesh: This bird is fairly common in **Uttar Pradesh**, particularly widespread in the north-west and central parts of the state which have various natural wetlands and low lying areas. Bundelkhand and the Vindhyan range have many manmade waterbodies. These waterbodies also sustain a fairly large number of waterbirds. There are several records throughout the state where its nesting is reported. For example, a congregation of more than 300 birds with hectic nesting activity was reported from **Sur Sarovar** WLS in August 2011 in the company of Black-headed Ibis, cormorants, egrets, herons, Eurasian Spoonbills, and Asian Openbill. Neeraj Mishra in September 2013 observed two Oriental Darters with two chicks in Kanpur Zoo, where a natural lake and vegetation around it offer ideal conditions for nesting and rearing these birds.

Although found all over **Uttar Pradesh** in suitable waterbodies, it has been specifically recorded from the following IBAs/PAs: **Lakh-Bahosi** Bird Sanctuary (Farrukhabad), **Narora** (Bulandshahr, Badaun), **Nawabganj** Bird Sanctuary (Unnao), **Patna** Bird Sanctuary (Etah), **Samaspur** Bird Sanctuary (Raebareli), **Sheikha** Jheel (Aligarh), **Sur Sarovar** Bird Sanctuary (Agra), **Bakhira** Jheel (Sant Kabir Nagar), **Okhla** Barrage and **Ramgarh** taal (Gorakhpur).

During a two-year avian influenza surveillance from May 2008 to March 2010, Rahmani *et al*. (2010) found it in the following wetlands (maximum numbers and date given in brackets): **Sheikha** (36 on January 11, 2009), **Aama-Khera** (8 on May 12, 2008), and **Daupur** (2 on February 4, 2009) in **Aligarh** district; **Patna Jheel** in **Etah** (3 on February 7, 2009); **Sur Sarovar** in **Agra** (48 on June 10, 2009); **Lakh-Bahosi** in **Farrukhabad** (82 on September 1, 2009); **Saman** (2 on September 28, 2009), **Sauj** (2 on January 22, 2010), **Kurra** (10 in January 8, 2010) in Mainpuri.

It was found in 20 out of 30 wetlands taken up for restoration in Sitapur district. More than 100 birds were found resting on a mound at **Jyotishah Alampur**, one of the potential wetlands in the district (Kumar & Srivastava 2011). Amit Mishra (*pers. comm.* 2011) spotted more than 50 birds on the right bank of **Girwa** river in **Katerniaghat** WLS during October 2008. Besides, the bird is common in **Sandi** WLS, **Surha Taal** WLS, **Vijay Sagar** WLS to name a few, and present in almost all reservoirs and barrages all over the state. It is recorded from some areas in western Uttar Pradesh, especially around **Hastinapur** WLS, **Bijnor** Barrage and **Okhla** Wildlife Sanctuary.

Rajat Bhargava and Nikhil Shende of BNHS during their bird surveys of the Terai from November 2013 to March 2014 found solitary birds on the following wetlands: Bhinga Range, **Shravasti** Forest Division (December 30, 2013); Chittorgarh dam, Rampur Range, **Soheldev** WLS (December 29, 2013); Singhrana Taal, North Chowk, **Sohagi Barwa** WLS (February 7, 2014); and Dudhwa Range, **Dudhwa** NP (March 14, 2014).

Ecology: It generally occurs singly or in small discrete groups, each one hunting fish independently. In good hunting grounds, up to 100 are seen, solitarily or in small groups. It is an expert diver and feeds almost exclusively on fish caught by its stiletto-shaped bill. It often swims with only the neck above water: the long neck and pointed bill give it the appearance of a snake, hence its popular name Snake Bird.

It nests colonially with egrets, storks, and herons, on thorny trees, generally half-submerged or near water. It makes a platform nest, sometimes very close to other nests, and lays 3–6 eggs which become soiled as incubation progresses. Chicks are blind and naked but soon develop white down feathers which may persist even when almost fledged. Some of the best breeding colonies of Darters can be seen in Sur Sarovar and Nawabganj Bird Sanctuaries.

Threats: Main threat to this and all piscivorous species is from excessive fishing all over its range, particularly in north India where fishing is very intensive. Pollution and spread of invasive species such as Water Hyacinth *Eichhornea crassipes* and *Ipomea carnea* are other problems. Hunting and disturbance at nesting colonies are not such a problem, at least in most parts of India as it is a protected species and traditionally protected by local communities along with Grey Pelican, Painted Stork, egrets and cormorants. Nevertheless, some tribals such as Gulgulawas and Kurmi-Baheliyas hunt it regularly for food.

Conservation measures underway: In India, it is listed in Schedule IV of the Wildlife (Protection) Act, 1972 and its hunting and disturbance are totally prohibited. Its trade is also banned. It occurs in a number of PAs/IBAs. The list is too long to give here. It also gets protection in the breeding colonies of Grey Pelican and Painted Stork which are traditionally protected by local villagers, and also in many temple tanks, where fishing and hunting are not allowed.

RECOMMENDATIONS

(1) Proper survey to determine its actual status in the state.

(2) Study on its ecology and habitat requirements.

(3) Regular monitoring (every 2–3 years) of major nesting sites to know the population trend.

(4) Involvement of local people for protection of its nesting sites, and if necessary augmentation of fish resources in some village ponds to improve its breeding success.

(5) Strict prohibition on fishing with very small mesh-size nets (zero-net fishing).

(6) Strong directive from the Forest Department and its flying squads to stop and nab poachers hunting birds in heronries.

Laggar Falcon

Falco jugger Gray 1834

DHRITIMAN MUKHERJEE

Despite its wide distribution in the Indian subcontinent, it is a poorly known species. It is supposed to be undergoing a moderately rapid population decline, owing to both pesticide use and incidental capture by trappers targeting the Saker Falcon *Falco cherrug* (BirdLife International 2013), hence it is listed as Near Threatened. If more accurate surveys prove that it is under greater threat, it may be uplisted to Vulnerable.

Field Characters: A large falcon (43–46 cm) with a white forehead and narrow supercilium above a black eyeline, and a long, narrow moustachial stripe running down from in front of and below the eyes. The crown is pale rufous with variable black streaks. Below it is white from chin to belly, with longitudinal light brown drops, in some individuals very faint on breast and belly, darker on flanks and thighs. Sexes are alike, but female is larger. Juvenile darker brown above with dark head and a broader moustache, and very heavily marked, almost uniformly dark, on breast, belly, and thighs. Immatures are even more heavily marked. In adults, legs and feet are yellow, with black claws, while juveniles have pale grey or greenish grey. For a detailed plumage description, see Ali & Ripley (1987), Rasmussen & Anderton (2005), and Naoroji (2007).

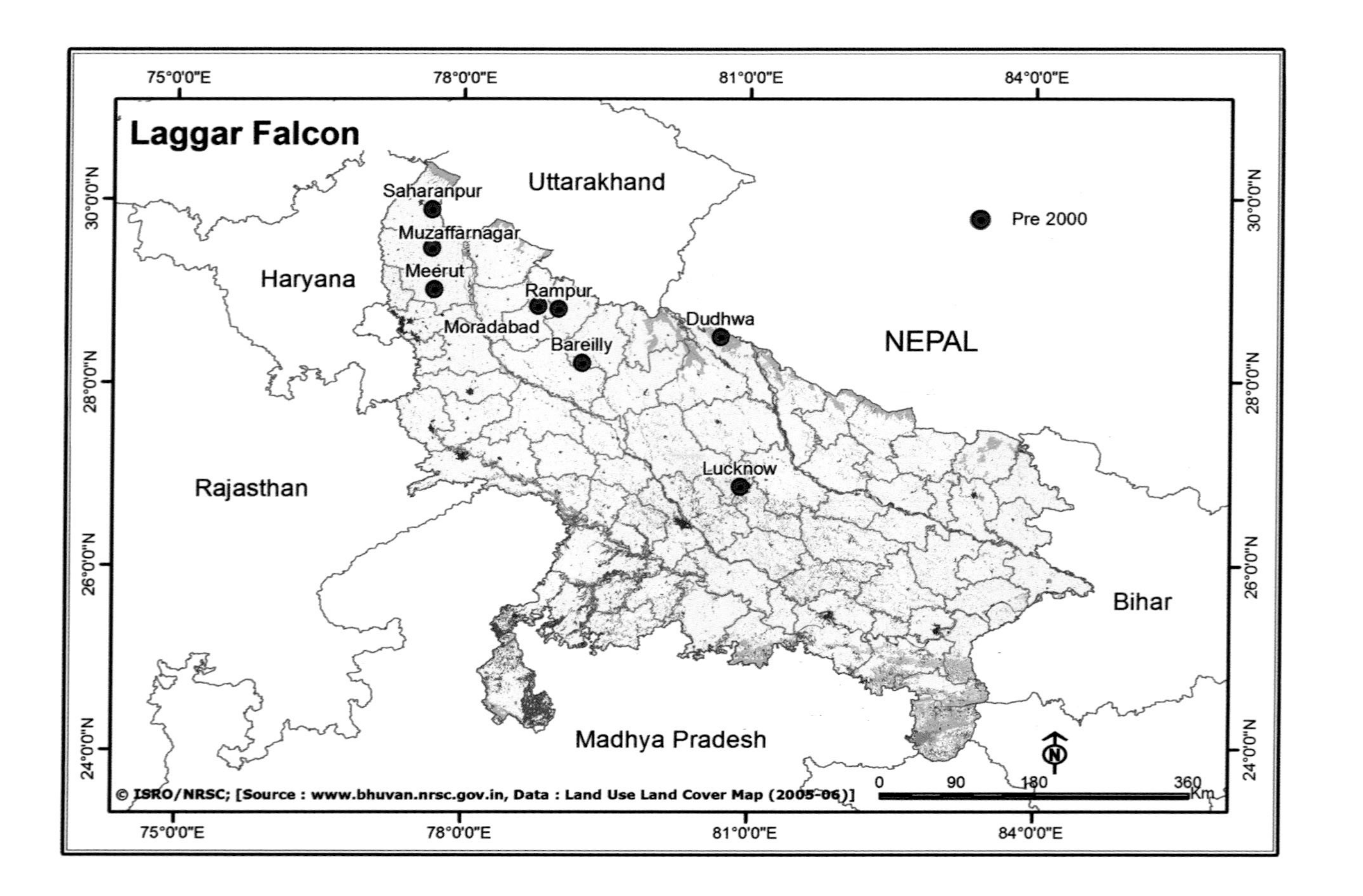
Laggar Falcon
Uttarakhand
Haryana
Rajasthan
Madhya Pradesh
Bihar
NEPAL
Pre 2000
Saharanpur
Muzaffarnagar
Meerut
Rampur
Moradabad
Bareilly
Dudhwa
Lucknow
75°0'0"E
78°0'0"E
81°0'0"E
84°0'0"E
30°0'0"N
28°0'0"N
26°0'0"N
24°0'0"N
0 90 180 360 Km
© ISRO/NRSC; [Source : www.bhuvan.nrsc.gov.in, Data : Land Use Land Cover Map (2005-06)]

ASAD R. RAHMANI

Laggar Falcon frequently occupies nests of other raptors and crows, even those located on pylons

Distribution: The Laggar Falcon is widely distributed in the Indian subcontinent, except in Sri Lanka. It is found locally in southern Afghanistan and possibly southeast Iran (Ferguson-Lees & Christie 2001). Its main population is found in northwest India, and smaller numbers in south India. In his monumental book *Birds of Prey of the Indian Subcontinent*, Rishad Naoroji (2007) has shown it as a rare resident in the whole of Uttar Pradesh. Javed & Rahmani (1998), during their study from 1991 to 1994 in Dudhwa National Park, did not observe it in the Park, perhaps as it is too dense and wet for the liking of this species.

In western UP, the Laggar Falcon was previously sighted regularly (in fact trapped) from Meerut (Manglore and Purkazi areas), **Muzaffarnagar**, **Saharanpur**, Rampur, **Moradabad**, **Bareilly**, and **Lucknow** districts, but there is no recent record for the last five years or more (Rajat Bhargava *pers. obs.* 2013). Quite recently, one specimen was trapped in Meerut in the first week of December 2013.

Ecology: In India, it is partial to drier biotopes, from sea level to 1,000 msl. It is usually seen perched on poles or on top of large bushes, generally solitary, but sometimes in pairs (in the breeding season). Naoroji (2007) has reported six birds together, but this could have been a family group. Preys on small birds, while

Blue Rock Pigeon *Columba livia* is regularly preyed upon in urban areas. Also feeds on flying insects (locusts), Spiny-tailed Lizard *Uromastix hardwickii*, and other reptiles. It does not build a nest but pirates or uses old nests of crows, kites, and even vultures. Usually three or four eggs are laid, and incubated mostly by the female, but both sexes feed the chicks. For details of behaviour and ecology, see Naoroji (2007).

Threats: Laggar Falcon is mainly threatened by increasing urbanisation, industrialisation, and expansion of agriculture in desert areas, particularly in the Indian Thar. Many traditional nesting sites have been abandoned due to increase in human population, disturbance, and traffic noise (Naoroji 2007).

The Blue Rock Pigeon is commonly traded in the Indian bird market, where on an average 400–500 trappers go out to catch it each day using a clap-trap with live pigeons tied as bait. Falcons in the vicinity get caught as a by-catch of the pigeon trade. The Laggar Falcon is often caught in this manner and sold for falconry, though it is not a much preferred species for falconry in India (Ahmed & Rahmani 1996, Ahmed 1997, Ahmed *et al.* 1997). Earlier, some were smuggled out via Nepal to Pakistan and finally to the Middle East, but presently due to the increased risk of being caught, this species is less targeted by smugglers, compared to the Peregrine and Saker Falcons (Ahmed 2012). It is also used as a decoy to trap Saker *Falco cherrug* and other larger falcons. It is caught comparatively easily employing a *Bal Chhatri* or pigeon harness.

Conservation measures underway: Like all birds of prey, it is legally protected and listed in Schedule I of the Indian Wildlife (Protection) Act, 1972. It is found in some PAs/IBAs of India (Islam & Rahmani 2004).

RECOMMENDATIONS

As its status and distribution is not well known in Uttar Pradesh, we recommend a thorough survey of this species, along with other raptors in the whole state. Our recommendations are as follows:

(1) Survey to assess the size of the population.

(2) Regularly monitor the population at selected sites across its range.

(3) Restrict the use of pesticides and make local people aware of the toxic effects of pesticides on the local wildlife.

(4) Enforce the legal protection afforded to the Saker Falcon, to benefit this species as well.

(5) Determine the level of capture of this species and its effect on population levels.

(6) Conduct extension education programmes to discourage falcon trapping.

Lesser Fish-eagle

Ichthyophaga humilis (Müller & Schlegel 1841)

RISHAD NAOROJI

According to BirdLife International (2013), the Lesser Fish-eagle is thought to be undergoing a moderate population reduction owing to forest degradation, overfishing, and perhaps especially pollution. It is consequently classified as Near Threatened.

Field Characters: Head and neck brownish grey to grey with black shaft-streaks, but otherwise brown above with black primaries; breast brown and clearly demarcated from white belly, thighs, and crissums (Ferguson-Lees & Christie 2001). According to J. Praveen (*pers. comm*. 2010), it has brown uppertail-coverts as against the white of the Grey-headed Fish-eagle. This is a very good field identification character, separating it from the Grey-headed Fish-eagle. The Lesser also differs from the larger Grey-headed Fish-eagle by the dusky base of its tail (*vs* white), wing tips nearly reach the tail tip, and seven primaries easily seen in flight (*vs* eight). Bill strong and small head on longish neck. Sexes similar but female usually larger. Juvenile with browner head and neck, paler brown above, white belly, and indistinctly streaked whitish on neck and breast (Ferguson-Lees & Christie 2001).

For more differences between the Lesser Fish-eagle and Grey-headed Fish-eagle, see Lethaby (2005).

Distribution: The Lesser Fish-eagle was earlier reported only from north and northeastern India. However, some birds were photographed, first by Vijay Cavale in December 2003, on the Kaveri river at Galibore, Karnataka, and later seen by many others. Outside India, it is found in Nepal, Bangladesh, Bhutan, Myanmar, Thailand, Cambodia, Laos, Vietnam, Malaysia, Indonesia, Brunei, and China.

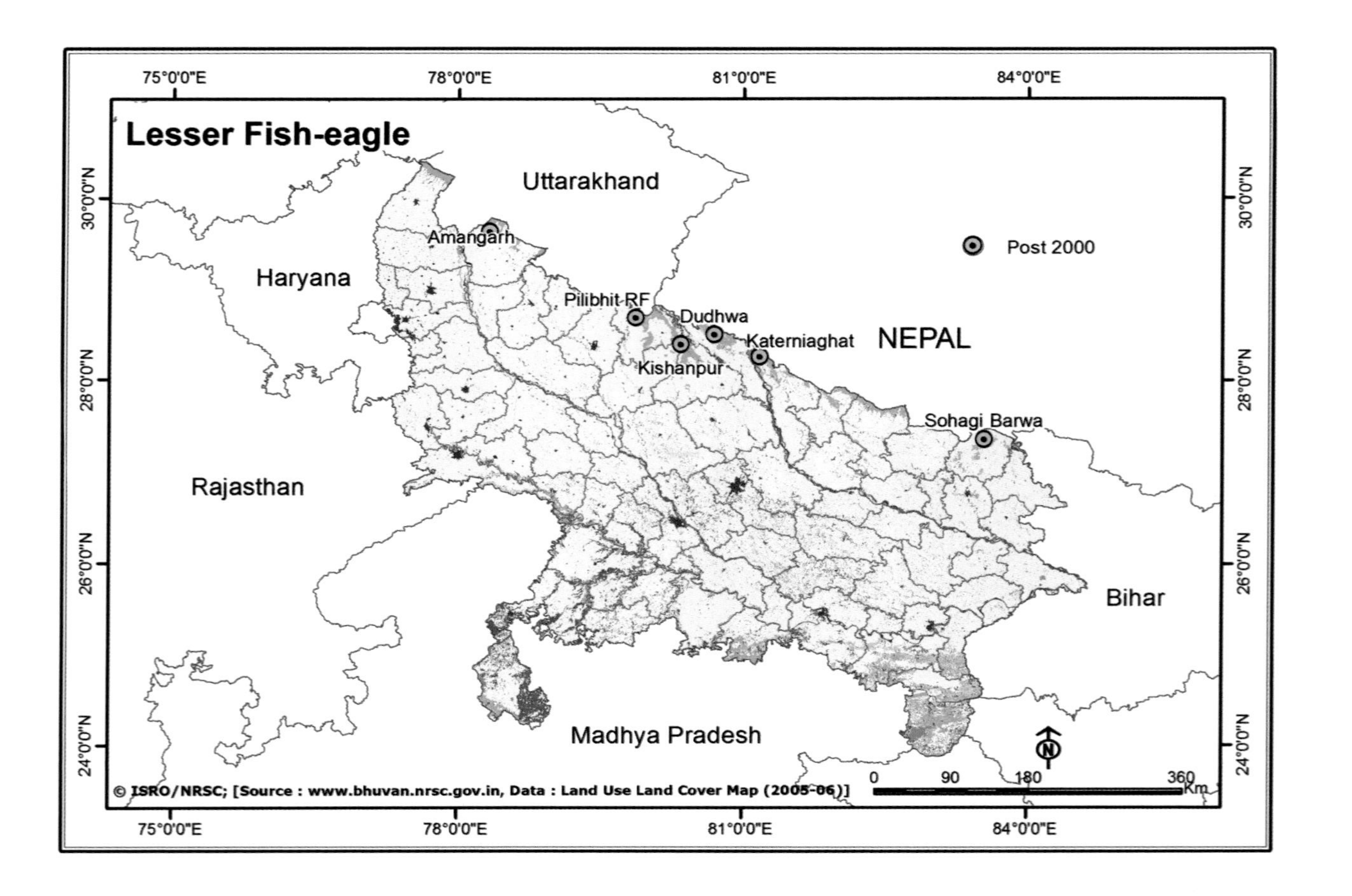

Lesser Fish-eagle
Uttarakhand
Haryana
Rajasthan
Madhya Pradesh
Bihar
NEPAL
Post 2000
Amangarh
Pilibhit RF
Dudhwa
Kishanpur
Katerniaghat
Sohagi Barwa
75°0'0"E
78°0'0"E
81°0'0"E
84°0'0"E
30°0'0"N
28°0'0"N
26°0'0"N
24°0'0"N
0
90
180
360
Km
© ISRO/NRSC; [Source : www.bhuvan.nrsc.gov.in, Data : Land Use Land Cover Map (2005-06)]

In **Uttar Pradesh**, it is reported from **Dudhwa** (Javed & Rahmani 1998) and it is likely to occur in Katerniaghat and Kishanpur, and maybe North Pilibhit Reserve Forests. In Bijnor district, it is said to occur in the **Amangarh** part of Corbett Tiger Reserve and **Najibabad** Forest Division.

Rajat Bhargava and Nikhil Shende of BNHS during their bird surveys in the Terai on February 7, 2014 found one in Singhrana Taal, North Chowk, **Sohagi Barwa** WLS.

Ecology: It frequents large forested rivers and wetlands in the lowlands and foothills up to 2,400 msl, but usually below 1,000 msl. It also occurs at sea level, e.g., in Sulawesi (del Hoyo *et al*. 1994). Its breeding ecology has been studied by Naoroji (2007) at **Corbett** NP. It prefers clear rapid forest streams in the lower Himalaya (*bhabhar* and the Terai), large jheels and reservoirs, preferably surrounded by forest. However, sometimes it is seen on open stretches of rivers, looking for fish, its favourite food. Said to be destructive to trout fisheries. It hunts from regular waterside perches, usually from bare branches or mid-stream rocks, dropping to snatch prey at or near the surface (Ferguson-Lees & Christie 2001).

Threats: Loss of forest habitat along rivers, siltation, overfishing, and increasing human disturbance in waterways are causing widespread decline in this species. Naoroji (1997) studied the effects of pesticide contamination in Corbett TR from 1991 to 1996. Eggs from seven of the nests monitored during this period did not hatch, while eggs in three nests hatched, but the young ones were either found dead in the nest, or had disappeared within a week of hatching. A number of organochlorine pesticides (DDT, dieldrin) were detected in the egg shells. These contaminants could have been passed on only through the prey base. In the higher reaches of the Ramganga river, which passes through Corbett TR, pesticides are heavily used on densely populated hillsides, the contaminants probably running off into the water system. Pesticide contamination, coupled with overfishing, damming of rivers, and destruction of riverine forests, are major threats to this species.

Conservation measures underway: Like all birds of prey, it is legally protected and listed in Schedule I of the Indian Wildlife (Protection) Act, 1972. It is also listed in CITES Appendix II.

RECOMMENDATIONS

(1) As it was earlier confused with the Grey-headed Fish-eagle, a proper survey of this species should be conducted to know its exact distribution in the state.

(2) Studies on its ecology and behaviour and exact habitat requirements should be initiated in India.

(3) Impact of pesticides on its food chain should be studied in the Terai of UP.

(4) Enforce control on destructive fishing such as dynamiting and small mesh-size (zero) nets in rivers and streams.

(5) Trees holding nests and large trees overlooking water sources should be specially protected from any form of destruction.

(6) Regular monitoring of its population should be initiated in Uttar Pradesh.

Grey-headed Fish-eagle

Ichthyophaga ichthyaetus (Horsfield 1821)

SANJAY KUMAR

BirdLife International (2013) justifies its inclusion in Near Threatened category as this species is thought to be undergoing a moderately rapid population reduction owing to habitat degradation, pollution, and overfishing. Although widespread, it is now only locally common. Uttar Pradesh is one of its strongholds in India.

Field Characters: A medium-sized raptor *c.* 69 to 74 cm, with grey head, neck, nape, and breast, merging with the paler brown of the mid-belly. Abdomen, flanks, and tail are white. Upperparts are brown, darker on the wings, turning to blackish on the quill tips. Terminal tail band is dark brown, and particularly visible in flight. Sexes are alike. Juvenile streaked overall, except on belly and vent, with white underwings and lightly barred flight feathers and tail.

Distribution: The Grey-headed Fish-eagle has a discontinuous distribution in India, perhaps due to its specialised habitat requirement of comparatively sluggish rivers and streams flowing through undisturbed forests. It is resident in the north and northeast Indian plains, Narmada river system, and the Western Ghats. A small population survives in the Gir forests of Gujarat. Outside India, it is found in the lowlands of Nepal, Bangladesh, Sri Lanka, Myanmar, the whole of southeast Asia, even up to Philippine Is. Nowhere is it abundant, and it has disappeared from many rivers due to disturbance and overfishing.

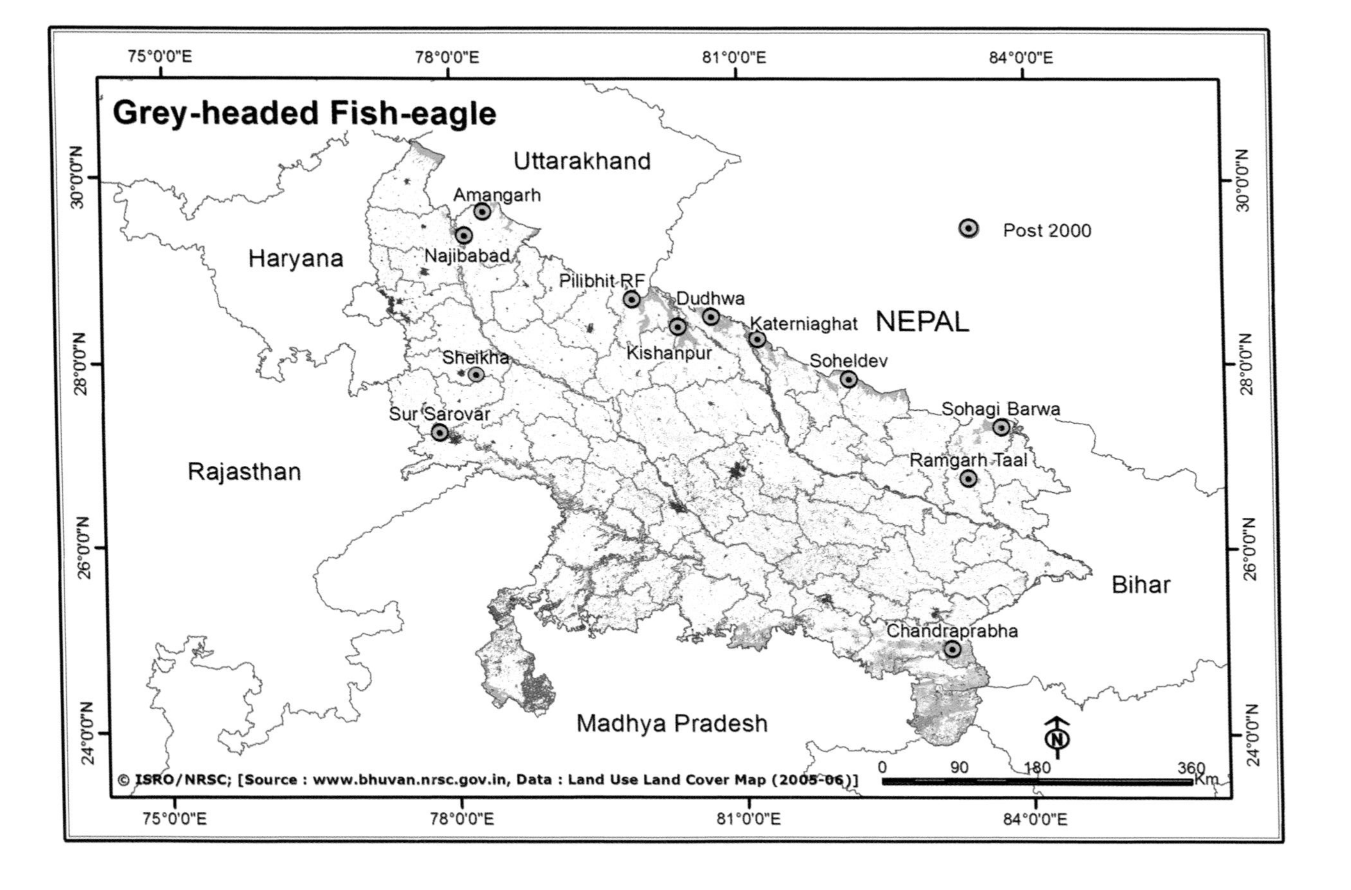

Grey-headed Fish-eagle
Uttarakhand
Haryana
Rajasthan
NEPAL
Bihar
Madhya Pradesh
Post 2000
Amangarh
Najibabad
Pilibhit-RF
Dudhwa
Kishanpur
Katerniaghat
Soheldev
Sohagi Barwa
Ramgarh Taal
Chandraprabha
Sheikha
Sur Sarovar
75°0'0"E
78°0'0"E
81°0'0"E
84°0'0"E
30°0'0"N
28°0'0"N
26°0'0"N
24°0'0"N
0
90
180
360
Km
© ISRO/NRSC; [Source : www.bhuvan.nrsc.gov.in, Data : Land Use Land Cover Map (2005-06)]

Records from Uttar Pradeah: The Grey-headed Fish-eagle is found in most of the protected areas in Uttar Pradesh. Due to its habitat preference, areas with sluggish rivers or streams are certain to have the presence of this bird. **Dudhwa** National Park, **Katerniaghat** WLS, and **Kishanpur** WLS have many such slow-moving rivers and streams, suitable habitats for this fish-eagle. Amit Mishra (*pers. comm.* 2010) has regularly spotted this bird near Bankey Taal in **Dudhwa** NP and on Bandha Road in **Katerniaghat** WLS from 2007 to 2013. During the breeding season, usually October to January, it is highly vocal and can be traced by its call. One of its regular sites is Jhadi Taal in **Kishanpur** WLS. It can also be seen in the **Pilibhit RF**, **Sohagi-Barwa**, **Soheldev**, **Najibabad** forest, and **Amangarh Reserve** which is a part of Corbett TR in Bijnor, Uttar Pradesh. It can easily be spotted in some wetlands surrounded by wooded areas, for example **Sheikha**, **Ramgarh Taal** (Gorakhpur), and **Sur Sarovar** (Agra). Rajat Bhargava and Nikhil Shende of BNHS during their bird surveys of the Terai found it in Rajhia Taal, Bankatwa Range, **Soheldev** WLS on December 29, 2014, and another bird in Dudhwa Range, **Dudhwa** NP on March 14, 2014. Ilyas & Khan (2006) have reported it from **Chandraprabha** Sanctuary, Mirzapur.

Ecology: The Grey-headed Fish-eagle is found near slow-moving rivers and streams, lakes, reservoirs, and tidal lagoons in wooded country, usually in lowlands but ascending locally to 1,525 m (BirdLife International 2013). It feeds exclusively on fish, sometimes very large ones, but during the breeding season it pursues birds and small mammals also. It will also consume dead fish (del Hoyo *et al.* 1994). It is highly territorial and makes a large nest with sticks on a tall tree. It breeds during winter from November to January in north India but even up to April in southern India. Studies on sharing of nest duties, nest-building, and incubation period are yet to be done (Naoroji 2007).

Threats: The most pertinent threats are the loss of undisturbed wetlands, overfishing, siltation, pollution, and general persecution. All the threats that apply to Indian wetland birds (drainage, pollution, overfishing, disturbance) affect this species also.

Conservation measures underway: Like all birds of prey, the Grey-headed Fish-eagle is listed in Schedule I of the Indian Wildlife (Protection) Act.

RECOMMENDATIONS

The measures suggested below apply to the Lesser Grey-headed Fish-eagle also:

(1) Studies on its ecology, behaviour, and exact habitat requirement should be initiated in India.
(2) Impact of pesticides on its food chain should be studied in the Terai in UP.
(3) Enforce control on destructive fishing, such as dynamiting and small mesh-size nets in rivers and streams.
(4) Trees holding nests and large trees overlooking water sources should be specially protected from any form of destruction.
(5) Regular monitoring of its population should be initiated in Uttar Pradesh.

Cinereous Vulture

Aegypius monachus (Linnaeus 1766)

ASAD R. RAHMANI

The Cinereous Vulture has a moderately small population which appears to be suffering an ongoing decline in its Asiatic strongholds, despite the fact that in parts of Europe its numbers are now increasing. Consequently, it qualifies as Near Threatened (BirdLife International 2013).

Field Characters: A huge, mostly black or dark brown vulture between 98 and 107 cm, with broad wings and short, often slightly wedge-shaped tail. Adults dark brown, while juveniles are blackish, with dark crown, ruff, and upper breast, contrasting with paler adults. An almost naked head, with massive bill, and crown, lores, and cheeks covered with black fur-like feathers and down. Bare skin of head and neck bluish grey. Sexes alike but female is larger (2–4%) and heavier (*c*. 7%) than male. It is one of the largest vultures found in India. In its wide distribution range, size increases from west to east, with the Mongolian and Chinese birds larger than the European birds (Ferguson-Lees & Christie 2001).

It has the typical unfeathered bald vulture head (actually covered in fine down) and dark markings around the eye, giving it a menacing skull-like appearance. The beak is brown, with a blue-grey cere, while legs and feet are grey.

Distribution: The Cinereous Vulture has a large range from southern Europe, North Morocco, Algeria, Sudan, the Middle East, Central Asia, up to Mongolia and east China. It has a small reintroduced population in France. It is resident except in those parts of its range where hard winters cause limited movement.

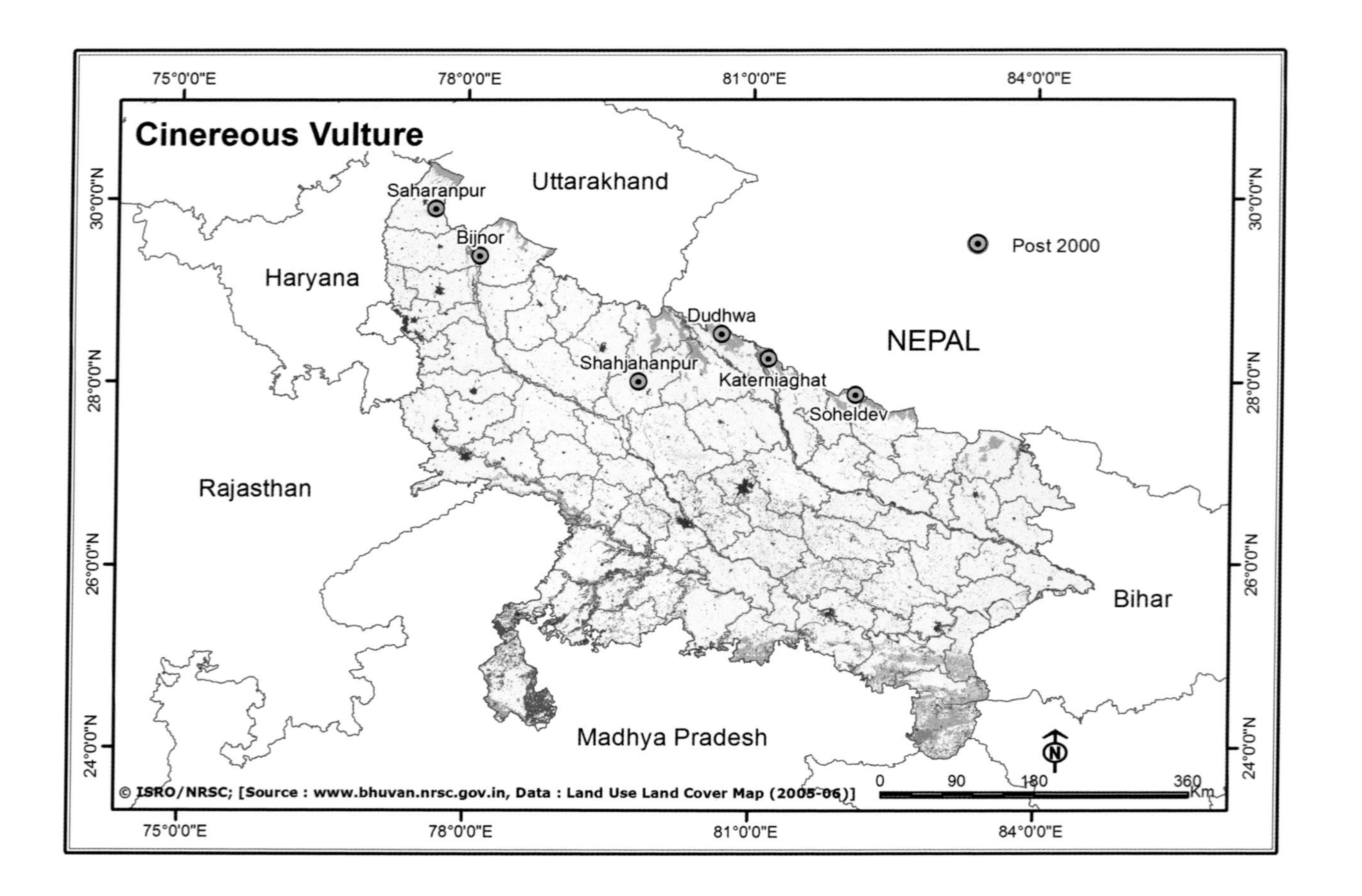
Cinereous Vulture
Uttarakhand
Saharanpur
Bijnor
Haryana
Post 2000
Dudhwa
NEPAL
Shahjahanpur
Katerniaghat
Soheldev
Rajasthan
Bihar
Madhya Pradesh
75°0'0"E
78°0'0"E
81°0'0"E
84°0'0"E
30°0'0"N
28°0'0"N
26°0'0"N
24°0'0"N
0 90 180 360 Km
© ISRO/NRSC; [Source : www.bhuvan.nrsc.gov.in, Data : Land Use Land Cover Map (2005-06)]

In India, it is a winter visitor, mainly in the Thar Desert and other dry biotopes. In **Uttar Pradesh**, we have some records: a single bird or two have been spotted in **Dudhwa** NP, **Katerniaghat**, and **Amangarh** Reserve (Bijnor) contiguous with Jhirna range of Corbett NP. Another sight record from Shahjahanpur range from Shivalik Forest Division in district **Saharanpur** is available. In February 2012, a bird was spotted perching on a leafless tree in sparse forest.

Rajat Bhargava and Nikhil Shinde of BNHS during their bird surveys in the Terai from November 2013 to March 2014 found three Cinereous Vultures scavenging in Poorvi Sohelwa Range in **Soheldev** WLS on December 28, 2013.

Ecology: In India, it is seen singly or in twos or threes, roosting early morning on large trees or sitting on sand dunes or mounds, commanding a grand view of the surroundings. In its breeding areas, it is also found in forested hills, as also alpine grasslands and steppes. It is generally a solitary nester, but sometimes it nests in very loose colonies or nuclei, with pairs 30 m to 2 km apart, and a record of 45 pairs with inter-nest distances of 100–400 m (Ferguson-Lees & Christie 2001). It feeds mainly on large mammal carcasses, but is also reported to feed on stranded turtles and large dead birds (Painted Stork). It dominates the jostling mixed rabble of vultures at carcasses, and is sometimes quite aggressive. It is equipped to tear open tough carcass skins, using its powerful bill.

Threats: BirdLife International (2013) has listed two major threats to Cinereous Vulture: direct mortality caused by humans (either accidentally or deliberately), and decreasing availability of food. Due to its large size and long life, it is also caught for zoos (Roberts 1991) and in China it is trapped or shot for feathers. In India, hunting is probably not a threat as this bird is left alone, and lack of food is also not a problem, particularly in the arid Thar Desert where the bulk of winter birds are found. However, diclofenac poisoning through livestock carcasses could be a major threat, though it is still not confirmed for this species.

Conservation measures underway: It is protected under the Indian Wildlife (Protection) Act, 1972. It is listed in Schedule IV of the Act.

RECOMMENDATIONS

Rahmani (2012) has given recommendations for its protection in India, which are also valid for Uttar Pradesh.

(1) In India mostly juveniles birds are recorded wintering. Juveniles of the species are known to wander long distances. Every winter a number of starving and dehydrated individuals are rescued. The birds recover within a few days when food and water are provided. There is a need to create awareness about the species, and people should be encouraged to send sick and injured birds to rescue centres. A lot of juveniles could be saved by providing water and care for a few days.

(2) Thorough post-mortem examination should be carried out of any dead bird found and the tissue should be checked for the presence of diclofenac.

(3) Regular surveys should be carried out throughout its known wintering range to determine its wintering population trends.

Pallid Harrier

Circus macrourus (Gmelin 1770)

YOGENDRA SHAH

Based on the steep population decline in Europe, BirdLife International (2013) has placed Pallid Harrier in the Near Threatened category of IUCN. Numbers in its Asiatic strongholds are thought to be more stable, and could be between 18,000 and 30,000.

Field Characters: A slender, lightly built, small (46–51 cm) grey and white harrier of grasslands, savannahs, and crop fields. Adult male almost white below, and has white wings with black wing tips. While sitting, the folded wings just reach the tail tip. Interestingly, its tail is not as long as other closely related harriers. Immature male may have rusty breast-band and juvenile facial markings. Female umber brown, with distinctive underwing pattern. For details of sex and age-related plumages, see Naoroji (2007) and for distinguishing it from the similar-looking Montagu's *C. pygargus* and Hen *C. cyaneus* Harriers, see Ali & Ripley (1987), Grimmett *et al*. (1999), Ferguson-Lees & Christie (2001), and Rasmussen & Anderton (2005).

Distribution: The Pallid Harrier breeds in eastern Europe, southern Russia, and Central Asia (including Turkey). In winter, it is found in the Indian subcontinent, the Middle East, eastern China, and mainly sub-Saharan Africa. It starts arriving in **India** by late August (mainly on passage through the Himalaya) and spreads out in the Subcontinent by October.

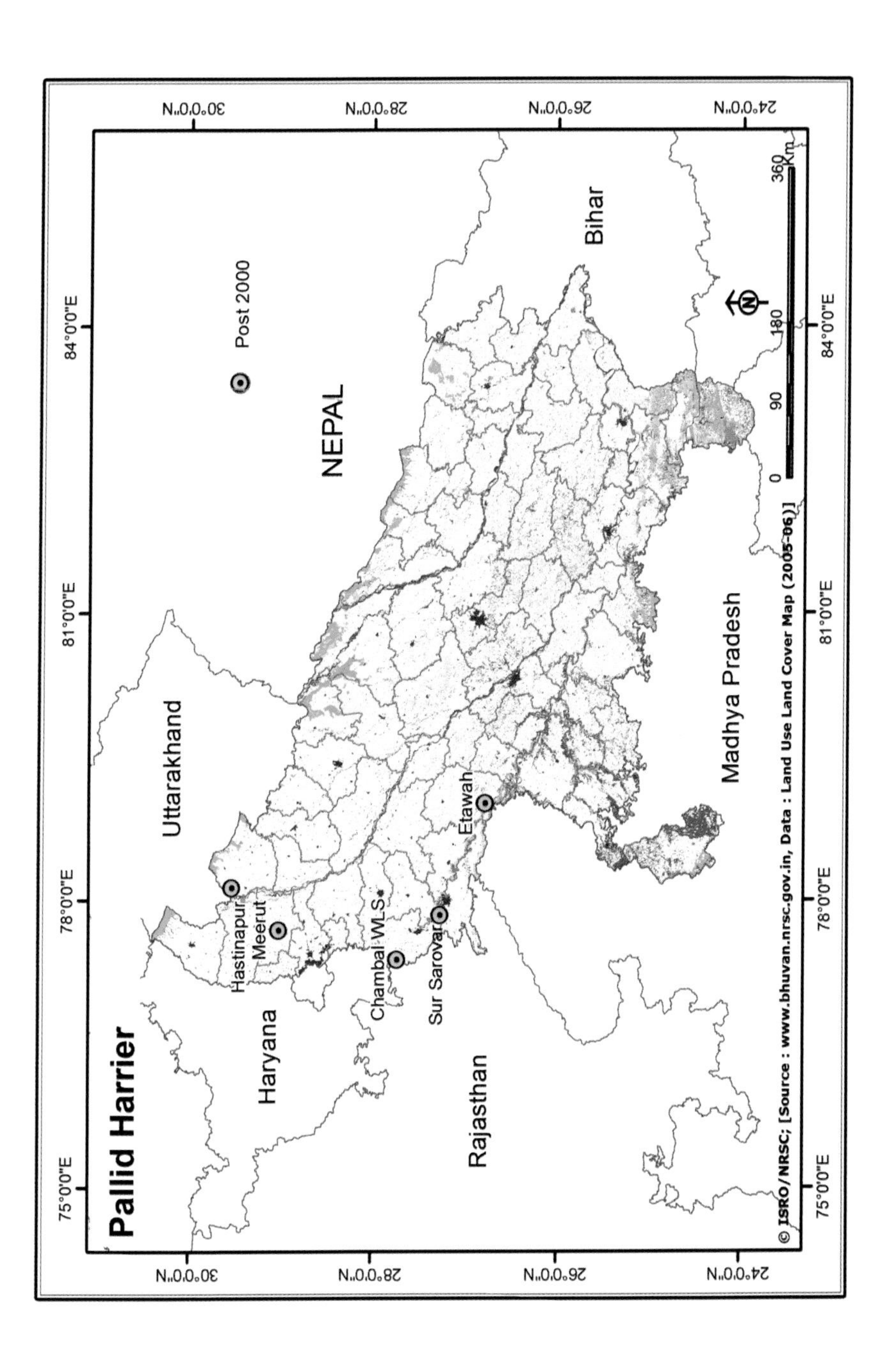

Pallid Harrier
Post 2000
NEPAL
Uttarakhand
Haryana
Rajasthan
Madhya Pradesh
Bihar
Hastinapur
Meerut
Chambal WLS
Sur Sarovar
Etawah
75°0'0"E
78°0'0"E
81°0'0"E
84°0'0"E
30°0'0"N
28°0'0"N
26°0'0"N
24°0'0"N
0
90
180
360
Km
© ISRO/NRSC; [Source : www.bhuvan.nrsc.gov.in, Data : Land Use Land Cover Map (2005-06)]

Records from Uttar Pradesh: Naoroji (2007) records the whole of Uttar Pradesh in its winter distribution range. However, Javed & Rahmani (1998) during their study on avifauna from 1991 to 1994 in Dudhwa NP did not report it: perhaps the Park's habitat is not suitable to this species. As it affects open, undulating high country, grassland, cultivation, desert, semi-arid and open scrub, we have to look for it outside forested protected areas in the state. It has been spotted in **Sur Sarovar** and **National Chambal Sanctuary** in the 8 km stretch between Nandgaon and Gudha. Based on the earlier work of Rai (1983), Bhargava (2012) has recently listed this species in his book *Birds of Meerut* as a passage migrant (with a question mark) in Meerut district, including **Hastinapur** WLS, but this requires more recent confirmatory records.

Ecology: In winter, it is found in open countryside, singly or in small groups, systematically skirting the ground for prey (largely small birds). In its winter quarters, it is attracted to grass fires which flush its prey, and locust swarms, and can be seen from sea level to 3,000 msl in the Himalaya and up to 4,000 msl in Africa (Ferguson-Lees & Christie 2001). It is usually silent in winter, but occasionally heard calling *keck-keck-keck-keck* at dusk, before finally settling to roost (Naoroji 2007). It roosts on the ground with other harriers. It breeds in wet grasslands close to small rivers and lakes, marshlands, semi-desert, steppe and forested steppe, and sometimes even in boreal forest and forested tundra zones, from sea level to 1,200 msl. Clutch size is four or five, and incubation lasts up to 30 days. The female alone incubates, with the male bringing food to her during incubation and early brooding. Breeding has not been reported in India.

Threats: The main threat to Pallid Harrier in the breeding areas is the destruction and degradation of grasslands through conversion of arable land to agriculture, burning of vegetation, intensive grazing of wet pastures, and clearance of shrubs and tall weeds. Destruction of grasslands and extensive use of pesticides and rodenticides are the major threats in its wintering grounds.

Conservation measures underway: Like all raptors, it is protected under the Indian Wildlife (Protection) Act, 1972. It is listed in Schedule I of the Act. Its hunting and trapping are strictly prohibited. It is listed in Appendix II of CITES, Annexure II of the Bonn and Bern Conventions and in Annexure I of the EU Birds Directive. It is not generally found in bird trade, mainly due to its restricted range and lack of specific demand. In India, it is found in many IBAs/PAs and grasslands.

RECOMMENDATIONS

Rahmani (2012) has given generic and specific recommendations. As it is marginal to Uttar Pradesh, our first priority would be to conduct a proper survey to find out its exact status in the state. These surveys should also result in identification of important grasslands where this bird winters. A telemetry-based study would also give good scientific results on its habitat use.

Eurasian Curlew
Numenius arquata (Linnaeus 1758)

VINAYAK YARDI

BirdLife International (2013) says that Eurasian Curlew remains common in many parts of its range, and determining population trends is problematic. Nevertheless, decline has been recorded in several key populations and overall a moderately rapid global decline is estimated. As a result, the species has been uplisted to Near Threatened.

Field Characters: A large (55–58 cm), pale sandy brown wader with long downcurved bill. It has mottled or scalloped brown plumage with whiter belly and undertail, and fine short black streaks on the underside. In flight it shows pointed whitish rump and barred tail as well as mottled whitish underwings. In flight, outer primaries are dark in contrast with whitish underwings. Flight slow and gull-like. Sexes alike. Juveniles are somewhat darker with finer black streaks on breast and very few on abdomen.

Distribution: The Eurasian Curlew breeds across the northern hemisphere and winters around the coasts of northwest Europe, the Mediterranean, Africa, the Middle East, the Indian Subcontinent, Southeast Asia, and east Asia. In **India**, it is a winter migrant to the entire country, particularly coastal areas, but also large jheels and rivers.

Records from Uttar Pradesh: In **Uttar Pradesh**, it is reported from many large wetlands, mainly on passage, but some birds stay throughout the winter. It has been reported from **Sur Sarovar**, **Bakhira**, **Narora,** and **National Chambal** Sanctuary.

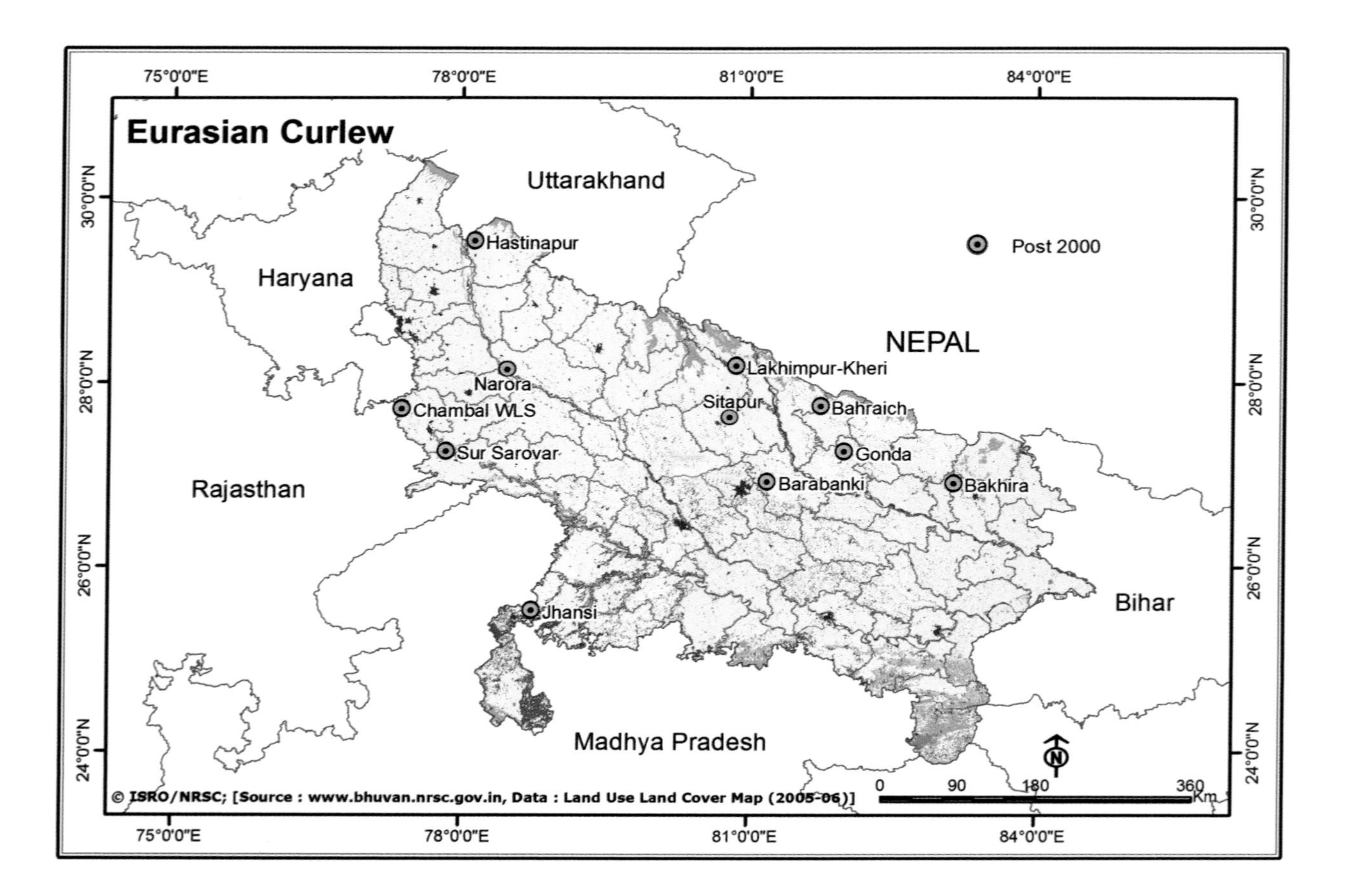
Eurasian Curlew
Uttarakhand
Haryana
Rajasthan
Madhya Pradesh
NEPAL
Bihar
Post 2000
Hastinapur
Narora
Chambal WLS
Sur Sarovar
Jhansi
Lakhimpur-Kheri
Sitapur
Bahraich
Gonda
Barabanki
Bakhira
0 90 180 360 Km
75°0'0"E
78°0'0"E
81°0'0"E
84°0'0"E
30°0'0"N
28°0'0"N
26°0'0"N
24°0'0"N
© ISRO/NRSC; [Source : www.bhuvan.nrsc.gov.in, Data : Land Use Land Cover Map (2005-06)]

We have one record from **Sur Sarovar** WLS where a solitary bird was sighted in October 2012 (Neeraj Mishra & Rishi Bajpayee *pers. comm.* 2013). Four birds were reported from **Narora** in February 2012 (P.D. Mishra & Raja Mandal *pers. comm.* 2012) and two birds in February 2013 (Amit Mishra *pers. comm.* 2013). Bhargava (2012) has listed this species in **Hastinapur** WLS, Meerut based on Rai's (1983) record. Birds are frequently seen on larger rivers such as Ganga, Yamuna, and Ghagra. We have records of Eurasian Curlew from the riverine islands and banks of river Sharda and Ghaghra passing through **Sitapur**, **Bahraich**, **Lakhimpur-Kheri**, **Barabanki**, and **Gonda districts**.

R.K. Sharma (*in litt.* 2010) conducted annual wildlife surveys in the **National Chambal Sanctuary** in Madhya Pradesh-Rajasthan-Uttar Pradesh, during February-March from Pali to Pachnada, a distance of 435 km. He counted the following numbers of Eurasian Curlew (year given in brackets): 7 (2003), 11 (2004), 5 (2005), 8 (2006), 16 (2007), 14 (2008), 8 (2009), and 10 (2010).

Ecology: The Eurasian Curlew is found singly or in small scattered flocks on coastal swamps, intertidal zones, beaches, edges of large rivers, jheels, and reservoirs, often in company with assorted waders and egrets. It feeds on annelid worms, aquatic insects and larvae, crustaceans, molluscs, spiders, berries, and seeds. It is also found feeding in damp grassland, sometimes far away from the coast. Small fish, amphibians, young lizards, young birds, and small rodents have been recorded in its diet (del Hoyo *et al.* 1996). Males are more likely to feed in inland grasslands than females. Some birds become territorial on wintering grounds, others feed gregariously (del Hoyo *et al.* 1996). It does not breed in India but in the temperate region, it has been found breeding on upland moors, peat bogs, swampy or dry heaths, fens, open grassy or boggy areas in forests, damp grasslands, meadows, dune valleys, and coastal marshlands (del Hoyo *et al.* 1996).

Threats: In Uttar Pradesh, the main threat is conversion of wetlands in agricultural fields, pollution, and trapping. Extensive trapping of all waders, including Eurasian Curlew, takes place in Bihar, **Uttar Pradesh**, Tamil Nadu, Andhra Pradesh, and certain parts of Gujarat (Ahmed 2002). The birds are sold directly to roadside hotels (*dhabas*) and sometimes brazenly advertised on the menu in some states!

Conservation measures underway: Like all waders, resident or migrant, it is protected under the Indian Wildlife (Protection) Act, 1972. It is listed in Schedule IV of the Act. This curlew occurs in many PAs and IBAs in India, and throughout its range features in several national monitoring schemes.

RECOMMENDATIONS

1) Protection of wetlands and periodic monitoring of Eurasian Curlew population through IBCN and AWC network.
2) Strict prevention of trapping and shooting.

River Tern

Sterna aurantia Gray 1831

SANJAY KUMAR

The River Tern has been uplisted to Near Threatened category because increasing human disturbance and dam construction projects are expected to drive a moderately rapid decline in its population over the next three generations (BirdLife International 2013).

Field Characters: A medium-sized tern, 38–43 cm long, with dark grey upperparts, white underparts, a forked tail with long flexible streamers, and long pointed wings. The bill is yellowish with black tip, legs red, and it has a black cap in breeding plumage. Sexes are similar, but juveniles have a brown head, brown-marked grey upperparts, grey breast sides, and white underparts.

Distribution: The River Tern is a common widespread resident in most of north and peninsular India. It breeds in summer on small islands formed in the backwaters of wetlands by the drying up of these waterbodies.

Records from Uttar Pradesh: Javed & Rahmani (1998) found it common in the wetlands of **Dudhwa** National Park. It is also fairly common in Jhadi Taal of **Kishanpur**, and Girwa river of **Katerniaghat Wildlife Sanctuaries**. It is reported from the following IBAs/PAs (Islam & Rahmani 2004): **Lakh-Bahosi** Bird Sanctuary (Farrukhabad), **Narora** (Bulandshahr, Badaun), **Saman** (Mainpuri), **Sandi** (Hardoi), **Bakhira** (Sant Kabir Nagar), **Hastinapur** (Meerut), **Kudaiyya** (Mainpuri), **Kurra** (Etah, Mainpuri), **National Chambal** WLS (Agra, Etawah), **Nawabganj** (Unnao), **Parvati Aranga** (Gonda), **Payagpur** (Bahraich), **Samaspur** (Raebareli), **Sauj** (Mainpuri), **Sur Sarovar** (Agra), **Surha Taal** (Ballia), and **Patna Jheel** (Etah). It can be easily spotted in the stretch of the River Ganga between Garhmukteshwar and **Narora** which is also a Ramsar site, Gomti river in the districts of **Barabanki** and **Jaunpur**, **Bijnor Barrage**, and near the *ghat*s of **Varanasi** and **Allahabad**.

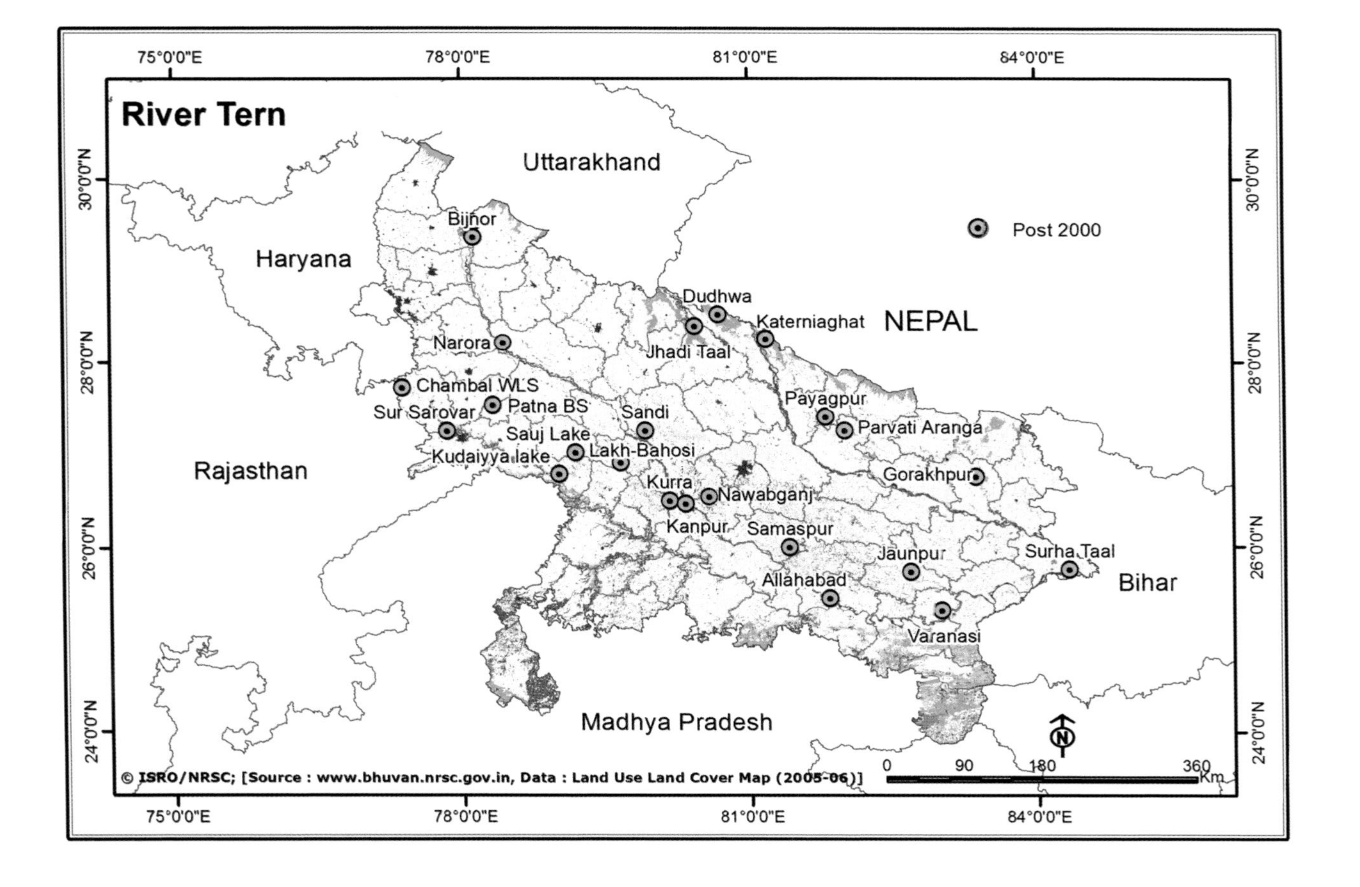
River Tern
Post 2000
NEPAL
Uttarakhand
Haryana
Rajasthan
Madhya Pradesh
Bihar
Bijnor
Narora
Chambal WLS
Sur Sarovar
Patna BS
Sauj Lake
Kudaiyya lake
Lakh-Bahosi
Sandi
Jhadi Taal
Dudhwa
Katerniaghat
Payagpur
Parvati Aranga
Gorakhpur
Kurra
Kanpur
Nawabganj
Samaspur
Allahabad
Jaunpur
Varanasi
Surha Taal
30°0'0"N
28°0'0"N
26°0'0"N
24°0'0"N
75°0'0"E
78°0'0"E
81°0'0"E
84°0'0"E
0
90
180
360
Km
N
© ISRO/NRSC; [Source : www.bhuvan.nrsc.gov.in, Data : Land Use Land Cover Map (2005-06)]

NEERAJ SRIVASTAV

Summer cultivation on river islands is one of the threats to the habitat of River Tern

The River Tern is not much described or studied in the state. From Narora, Amit Mishra (*pers. comm.* 2013) recorded more than 50 birds during a bird count near the barrage in February 2012. Chandan Prateek (*pers. comm.* 2013) too reported 15 birds from areas near **Rapti** river in **Gorakhpur** district in December 2012. More than 30 birds were recorded through collective sightings within 2 km on the banks of River Ganga in Bithoor, **Kanpur** in February 2012.

Ecology: This species inhabits rivers and freshwater lakes, also occurring rarely on estuaries. It feeds predominantly on insects and small fish. It breeds from early March to early May and breeding occurs mainly in colonies in less accessible areas such as islands and sandbanks in rivers (del Hoyo *et al*. 1996). It nests in a scrape on the ground, often on bare rock or sand, and lays three greenish grey to buff eggs, which are blotched and streaked with brown.

Threats: Nesting areas are vulnerable to flooding, predation, and other disturbance (del Hoyo *et al*. 1996). The numerous dam construction projects completed, underway, or planned may also threaten the species through changes in the flow regime and flooding of nest sites. Excessive fishing is another major threat. For example, four chicks of River Tern were found strangled in pieces of fishnets lying in nesting areas of River Tern at a breeding colony in the Bhima river in May 2013. The eggs of River Tern are lifted by villagers and they are also trampled upon by livestock which are brought to reservoirs to drink in the summer (Nandkishor Dudhe *pers. comm.* 2013).

RECOMMENDATIONS

(1) Carry out regular surveys to monitor the population throughout its range.

(2) Conduct education activities to help alleviate human pressures on river and lake habitats.

(3) Conduct satellite tracking to study its movement.

River Lapwing

Vanellus duvaucelli (Lesson 1826)

SANJAY KUMAR

BirdLife International (2013) has uplisted River Lapwing to Near Threatened as it is expected to undergo a moderately rapid population decline over the next three generations, owing to human pressures on riverine ecosystems and the construction of dams.

Field Characters: The River Lapwing is 29–32 cm long. It has a black crest, crown, face, bill, centre of throat, and legs, and grey-white neck sides and nape. The underparts are white with a black belly patch, and it has a grey-brown breast band. The back is brown, rump white, and the tail black. In flight, the black primaries, white underwings and upperwing secondaries, and brown upperwing-coverts are conspicuous. The call of the River Lapwing is a sharp *did-did-do-weet*, reminiscent of the call of the Red-wattled Lapwing, yet distinctive.

Distribution: The River Lapwing is resident throughout the large river systems of northern India and the foothills of Himalaya from western Himachal to Arunachal and the Assam valley; South Assam Hills (Meghalaya and south to Lushai Hills); Central India and northeastern Peninsula (Rasmussen & Anderton 2012). According to Ali & Ripley (1987), it is resident with some nomadic movement.

Records from Uttar Pradesh: It is found in all larger rivers and their tributaries, particularly the Ganga, Yamuna, Ghagra, Sharda, Chambal, Ramganga, Gomti, and Rapti. It can also be spotted in some wetlands connected with rivers like **Okhla** Barrage on the UP-Delhi border, and the 535 ha Ramgarh Taal inside **Gorakhpur** city. The lake is connected with a small channel to Rapti river.

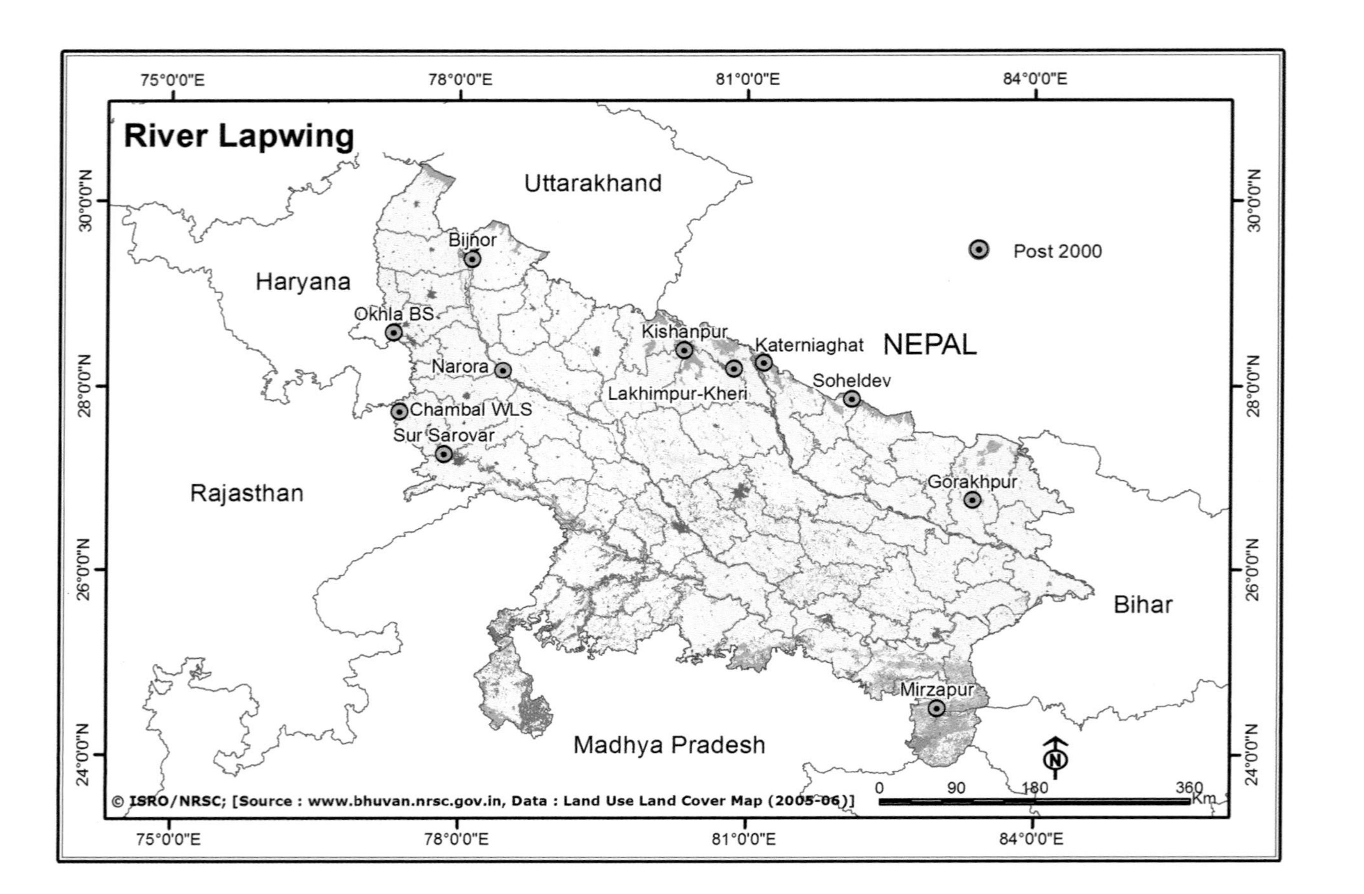

River Lapwing
Uttarakhand
Haryana
Rajasthan
NEPAL
Bihar
Madhya Pradesh
Post 2000
Bijnor
Okhla BS
Narora
Chambal WLS
Sur Sarovar
Kishanpur
Lakhimpur-Kheri
Kateriaghat
Soheldev
Gorakhpur
Mirzapur
0
90
180
360
Km
75°0'0"E
78°0'0"E
81°0'0"E
84°0'0"E
30°0'0"N
28°0'0"N
26°0'0"N
24°0'0"N
© ISRO/NRSC; [Source : www.bhuvan.nrsc.gov.in, Data : Land Use Land Cover Map (2005-06)]

In western Uttar Pradesh, the River Lapwing is seen throughout **Hastinapur** WLS (Bhargava 2012), **Garhmukteshwar**, and on the margins of **Bijnor Barrage**. During a wildlife survey conducted by Meerut Forest Department by boat from Hastinapur WLS to Jyotiba Phule Nagar in mid 2012, more than 15 individuals were seen, mainly solitary or in pairs on the dry sandbanks along the Ganga (Rajat Bhargava & Lalit Kumar Verma *pers. obs.*).

Katerniaghat and **National Chambal** Sanctuaries are two IBAs where this bird is resident and fairly common. It is found round the year except in the monsoon. In **Katerniaghat** WLS, its inconspicuous presence among pratincoles and other waders is seldom noticed. On the sandbars in the Girwa river, two or three pairs can be seen on sandy islands. Outside protected areas, a total of 50 birds (various sightings in 1 km collectively) were reported on the banks of River Ganga in **Narora** in January 2011. A flock of 10 birds was reported by Chandan Prateek (*pers. comm.* 2013) from **Gorakhpur** on the sandbanks of River Rapti in December 2012. Amit Mishra (*pers. comm.* 2013) counted 20 to 30 birds at Vindhyachal in **Mirzapur** district where they roost on the expanded sandbanks of Ganga river. He has observations for the last four years. Sightings were reported from Sharda river, **Kishanpur** WLS, **Lakhimpur-Kheri**, and Yamuna river, besides **Sur Sarovar**, **Agra**. Rajat Bhargava and Nikhil Shinde of BNHS during their bird surveys of the Terai from November 2013 to March 2014 found one pair on Rampur Dam, Poorvi Sohelwa, **Soheldev** WLS on December 27, 2014. These are only some of the known records.

As the bird is widely distributed in the rivers of Uttar Pradesh, it is difficult to say how many are present in the state.

NEERAJ SRIVASTAV

Overfishing, cultivation on river banks and islands, and sudden release of water from dams are major disturbances to River Lapwing, particularly during the breeding season from March to June

SANJAY KUMAR

River Lapwing and Indian Skimmer share the same habitat and face similar problems of human disturbances in their breeding areas

Ecology: Normally keeps singly or in pairs. In general, the ecology is very similar to that of other lapwings. The coloration is remarkably obliterative in its habitat of dry river sand and shingle. It feeds on insects, worms, and crustaceans. It breeds from March to June; the nest is a shallow scrape on exposed sand or shingle. Generally four eggs are laid (Ali & Ripley 1987).

Threats: Disturbance to breeding birds by humans, dogs, cats, and crows is the biggest danger to this and other species inhabiting large Indian rivers, as the ecology of rivers is under great pressure. Construction of dams and reservoirs, sudden release of water, depletion and diversion of water exposing the nests to ground predators, increasing watermelon cultivation on sandy river islands and consequently constant human presence, all these pressures may be working against the species. These threats need to be quantified in its whole range of distribution and effective site-specific measures need to be taken. River Lapwing is occasionally seen in bird markets when it is caught in the clap-traps laid for other waterbirds. Organized collection of sand for commercial trade and associated transport is a disturbance in some pockets for breeding birds.

RECOMMENDATIONS

We propose the following conservation measures:

(1) Proper baseline survey to know its status and distribution in the state.

(2) Protection of small river islands and sandspits in larger rivers where this bird breeds.

(3) Ringing and satellite tracking studies to monitor its movement and dispersal.

European Roller

Coracias garrulus Linnaeus 1758

SUNIL SINGHAL

According to BirdLife International (2013), European Roller has apparently undergone a moderately rapid decline across its global range and it was consequently uplisted from Least Concern to Near Threatened in 2005. Decline has been most pronounced in northern populations, and if similar trends are observed elsewhere in the species' range, it may warrant uplisting to Vulnerable. Its present global population could be between 100,000 and 500,000, but it is decreasing.

Field Characters: A pale roller with turquoise head and underparts, and tawny mantle. Wings blue with black flight feathers, clearly visible in flight. Large head and black bill, upper mandible hooked at the tip. Sexes similar, and the juvenile is a drab version of the adult.

Distribution: The European Roller has a huge breeding range in Europe, northern Africa (Morocco), the Middle East (Iraq, Iran), Central Asia, and northwest China. BirdLife International (2013) has shown that it is undergoing population decline in many areas, and overall European decline exceeded 30% in three generations (15 years). However, there is no evidence of any decline in Central Asia. Should these populations be shown to be declining, the species may warrant uplisting further to Vulnerable.

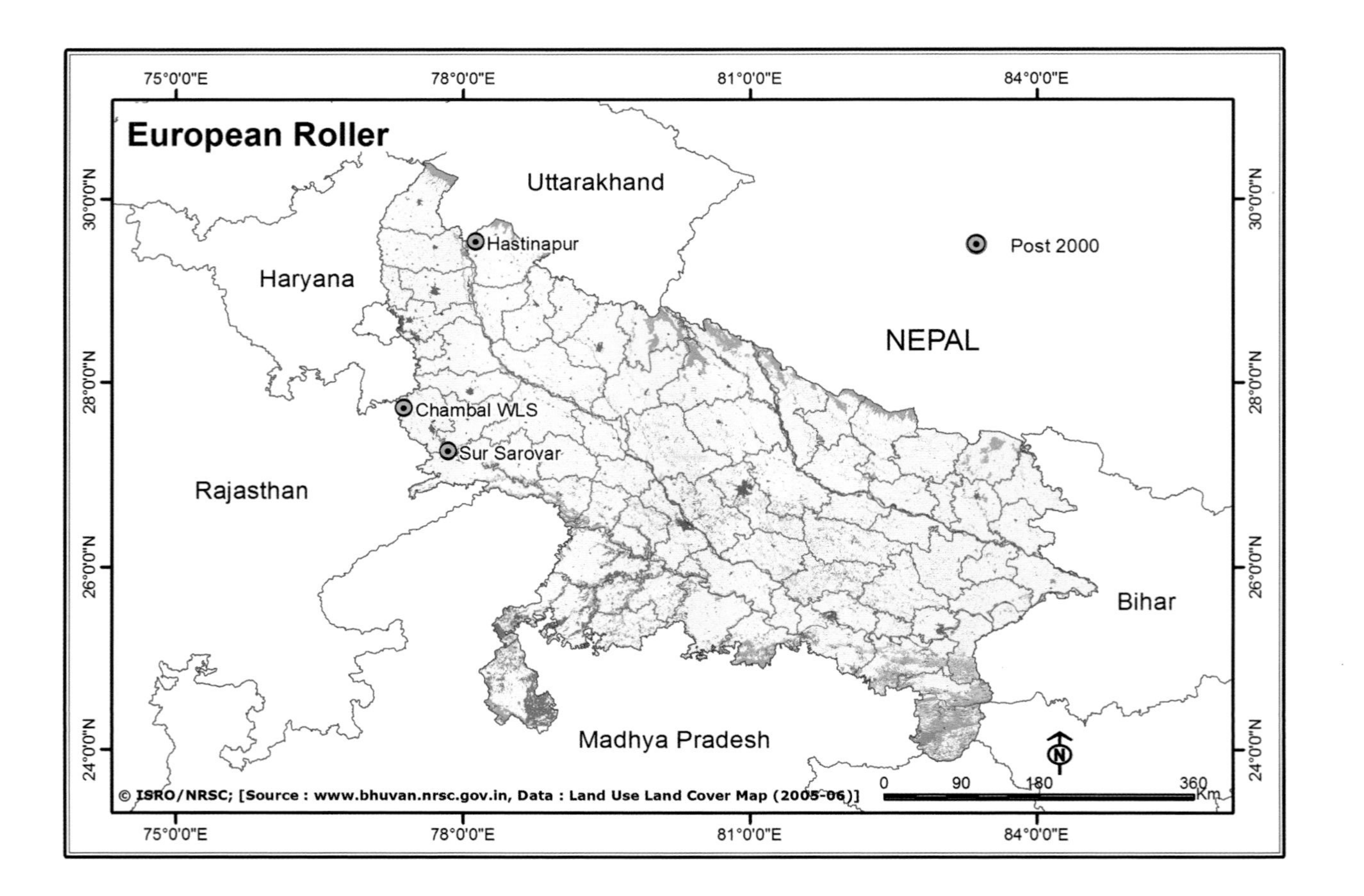
European Roller
Uttarakhand
Hastinapur
Post 2000
Haryana
NEPAL
Chambal WLS
Sur Sarovar
Rajasthan
Bihar
Madhya Pradesh
75°0'0"E
78°0'0"E
81°0'0"E
84°0'0"E
30°0'0"N
28°0'0"N
26°0'0"N
24°0'0"N
0 90 180 360 Km
© ISRO/NRSC; [Source : www.bhuvan.nrsc.gov.in, Data : Land Use Land Cover Map (2005-06)]

In **India**, the European Roller, also called Kashmir Roller, breeds only in the **Kashmir Valley**, otherwise it is a passage migrant, passing through northwest India from end August to September, while some individuals linger till October. It is seen mainly on passage in western **Uttar Pradesh**. It has been spotted at **Sur Sarovar** and **National Chambal** Sanctuaries along the Chambal ravines. It is regularly seen in Meerut district, including **Hastinapur** WLS, at the onset of winter each year in October and it probably stays for less than two to three weeks (Bhargava 2012).

Ecology: In the Kashmir Valley, it breeds in open wooded areas, parks, and cultivated areas, nesting in a natural tree hollow from 3 to 10 m above ground level, but "perhaps the most favoured site is a cavity in a sand-bank" (Bates & Lowther 1952). The clutch is four to five eggs, and chicks hatch by end June or early July (in Kashmir). Both parents incubate the eggs and rear the chicks. They forage in open cultivated areas, fallow fields, and open woodlands for large insects, ground beetles, lizards, frogs, small rodents, and small injured birds.

Threats: In **India**, the main threat to this species (and most insectivorous birds) is through direct pesticide poisoning or indirectly through decrease in insect supply. The Indian Roller or Neelkanth *Coracias benghalensis* is considered an incarnation of Lord Shiva and during the Hindu festival of Dassera sighting it is considered auspicious. In October, the Indian Roller is caught in large numbers for the release trade on Dassera day, when people pay to release this sacred bird. European Roller is a passage migrant at this time and gets caught because of its resemblance to the Indian Roller (Abrar Ahmad *pers. comm.* 2010). According to TRAFFIC India field surveys in several bird markets across Uttar Pradesh, it is estimated that a minimum of 500–600 rollers of both species are caught a week prior to Dassera each year, apart from the year round trapping for 'merit release' (Ahmed 2000), especially in cities such as Varanasi, Lucknow, Kanpur, Bareilly, Sitapur, Shahjahanpur, Moradabad, Meerut, and Agra, despite the ban on bird trapping in India since 1990. Rollers are caught by latex and mice method, and once they are released in urban areas they cannot survive due to loss of flight feathers while trapping or heaviness due to the glued feathers, combined with inter-species fights in captivity and starvation (Ahmed 1997, 2000, 2012). In certain villages, especially in eastern UP, rollers are caught for consumption by some traditional bird trapping tribes.

Conservation measures underway: It is protected under the Wildlife (Protection) Act, 1972. It is listed in Schedule IV of the Act. It is found in, or passes through, many PAs/IBAs.

RECOMMENDATIONS

In Uttar Pradesh, it is a marginal species seen in some areas during passage migration. However, as it is caught along with Indian Roller for religious rituals, there should be strict control on trapping, particularly during Dassera and other festivals.

Great Pied Hornbill

Buceros bicornis Linnaeus 1758

SANJAY KUMAR

Despite its very large range in South and Southeast Asia, the Great Pied Hornbill is patchily distributed in evergreen and moist deciduous forests, and occurs at low densities. It is likely to be declining moderately rapidly throughout its range, and is therefore considered Near Threatened by IUCN and BirdLife International (2013).

Field Characters: A large (130 cm), mainly pied hornbill with an enormous horn-shaped yellow-and-black bill, surmounted by a ponderous concave-topped casque, broadly U-shaped when viewed from the front (Ali & Ripley 1987). It has a black face, throat, mantle, and lower breast, buffy neck and nape, and white belly and vent. The white neck and wing-bars, and black band on the white tail, are diagnostic in flight. The casque is rectangular in cross section, double-pointed in front, round at the back, and concave or convex on top. It has bristles around the eyes that resemble eyelashes. Flight is often noisy as air rushes through the bases of the flight feathers, which are not covered with stiff coverts.

The Great Pied Hornbill is sexually dimorphic. The female is slightly smaller and has pink orbital skin, white iris, and red on the rear of the otherwise plain casque (Rasmussen & Anderton 2005). The male has a deep red iris with black orbital skin.

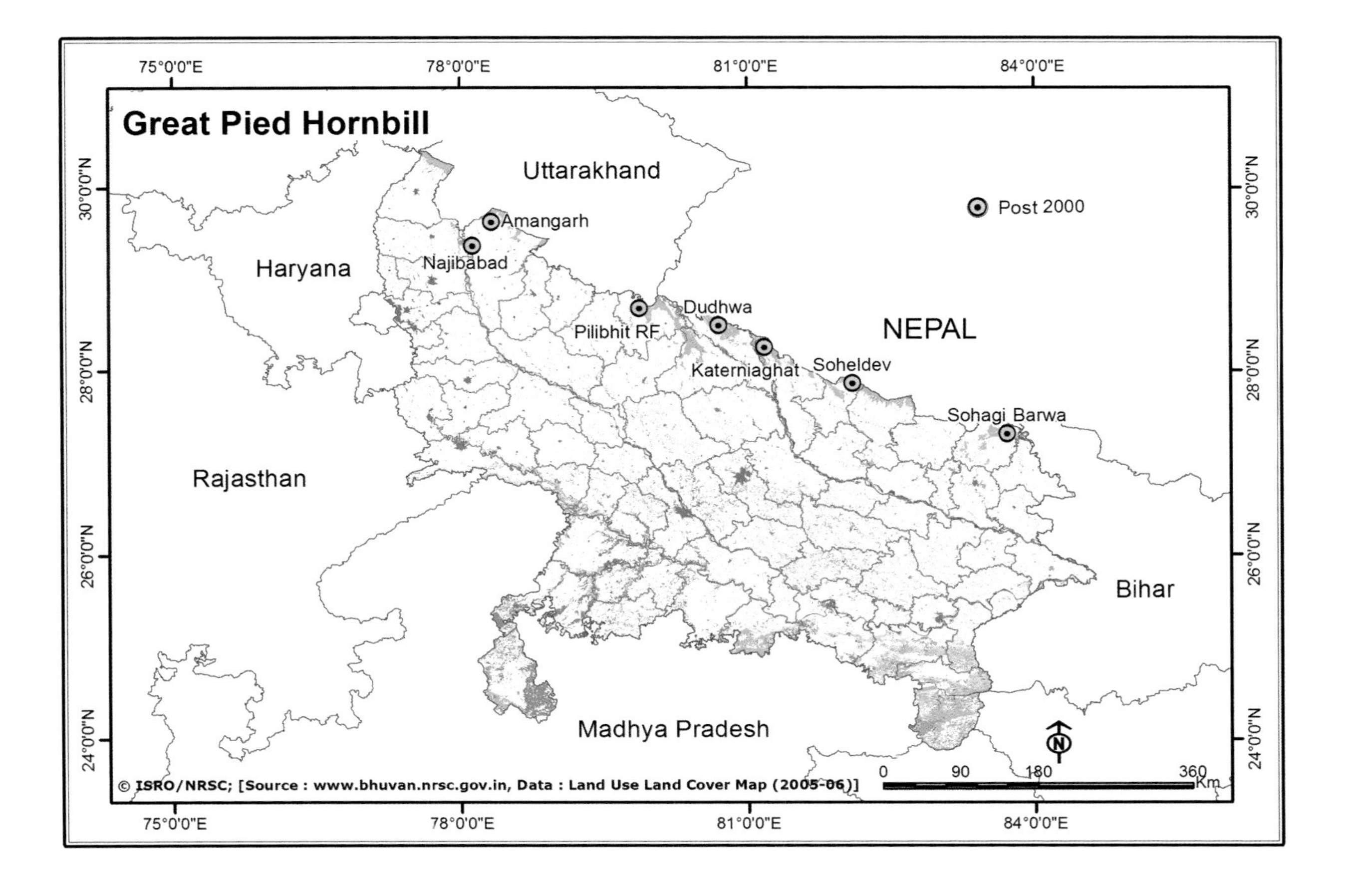

Great Pied Hornbill
Post 2000
NEPAL
Uttarakhand
Haryana
Rajasthan
Madhya Pradesh
Bihar
Amangarh
Najibabad
Pilibhit RF
Dudhwa
Katerniaghat
Soheldev
Sohagi Barwa
75°0'0"E
78°0'0"E
81°0'0"E
84°0'0"E
30°0'0"N
28°0'0"N
26°0'0"N
24°0'0"N
0
90
180
360
Km
© ISRO/NRSC; [Source : www.bhuvan.nrsc.gov.in, Data : Land Use Land Cover Map (2005-06)]

Distribution: It is found in evergreen and moist deciduous forests of South and Southeast Asia, up to 1,500 msl. In **India**, it is resident in the Himalaya from Uttarakhand to Arunachal Pradesh, Assam, Meghalaya, Orissa, and disjunctly in central India (thick forests, e.g., Bandhavgarh) and then in the Western Ghats. Outside India, it is found in Nepal, Bhutan, Bangladesh (rare), Myanmar, Thailand, Laos, Vietnam, Cambodia, southern China, Malaysia, and Indonesia. It is not abundant anywhere and generally declining, particularly in unprotected areas and logged forests (Kemp 1995, Datta 1998). Rahmani (2012) has collated the latest distribution records from India. In Uttar Pradesh, it has been seen in **Dudhwa** NP, **Katerniaghat** WLS, **Sohagi Barwa** WLS, **Suheldev** WLS, **Amangarh Reserve** , **Najibabad** forest, and **Pilibhit** forest. In **Katerniaghat** WLS, Nishangarha Range is a good place to find this bird. The best time to locate it in the thick *terai* forests is from March to May as this period coincides with their nesting season so the birds are more vocal.

Ecology: This species has been extensively researched. Here we give a brief description of its ecology and behaviour. This arboreal bird lives in small parties of three to five birds, but on fruiting trees or during roosting time, sometimes 30–40 birds are seen. They are predominantly frugivorous, although they are opportunistic and will prey on small mammals, reptiles, and birds. These long-lived large birds inhabit pristine forest with a mix of fruiting trees which fruit at different times of the year, which these birds know, ensuring a continuous food supply. Sometimes they can successfully nest in fairly disturbed and human-dominated landscape (Rahmani 2012). Great Pied Hornbills are monogamous. They are secondary cavity nesters and prefer the same nest over and over again as long as it is not disturbed. The length of the nesting cycle (from female nest entry and sealing to chick fledging) ranges from 110 to 129 days. The breeding season spans 20–22 weeks (March to August), though pre-breeding activities such as courtship and nest inspection are observed from January onwards. The females enter the nest cavities only in March. Females seal themselves in nest cavities using mainly their faeces that contain fig seeds, fruit pulp, and insect chitin.

Threats: Deforestation and commercial logging of large old trees are major problems, besides targeted hunting in some areas for food and traditional use of its large casque. Rahmani (2012) has described the types of threats in various parts of its range.

Great Pied Hornbill is a favourite trade species, primarily for zoos and supposed medicinal value. Its large size, omnivorous diet, rich coloration, and long lifespan make it an interesting avicultural subject and hence a victim of trade (Ahmed 2009, 2012). The bird is also poached for its beak, meat, fat, and feathers especially in northeast India. Traditional medicines prepared from hornbill fat are supposed to cure back pain and other ailments. Investigations carried out by TRAFFIC ndia reveal that in Uttar Pradesh hornbill chicks are first collected from

SANJAY KUMAR

Great Pied Hornbill survives in old-growth forest with sufficient numbers of Ficus trees to provide it food throughout the year

certain protected areas and primarily sent to Balrampur and Bahraich where they are hand-reared; from there they are sent to Lucknow, Allahabad, and Siwan (Bihar) for further sale to zoos and large private collections throughout India. It was revealed that this species is even smuggled to some Asian countries via Patna (Bihar) and Burdwan (West Bengal) (Ahmed 2009).

Conservation measures underway: It is protected under the Wildlife (Protection) Act, 1972 and listed in Schedule I. It is also listed in Appendix I of CITES. It is found in many PAs/IBAs.

RECOMMENDATIONS

(1) The most important action required is the protection of old growth forests.

(2) As we do not have data on this species in Uttar Pradesh, a proper survey should be conducted to identify its main areas of occurrence, where fig species could be maintained and enhanced.

(3) Strict prevention of collection and trade.

Rufous-rumped Grassbird

Graminicola bengalensis Jerdon 1863

JAMES EATON

Like most obligate grassland species, Rufous-rumped Grassbird may have suffered moderate to severe decline owing to the destruction or conversion of its tall wet grassland habitats. It is also difficult to detect. IUCN, based on the assessment by BirdLife International (2013), considers it Near Threatened.

Field Characters: A rufous-brown, moderate-sized grassbird (16 cm) with darkly-streaked crown, nape, and mantle; unstreaked rufous rump, and broad, dark brown, graduated tail with white tips. Chin, throat, breast, and upper belly are whitish, merging with its light rufous lower belly. In extremely worn plumage, white tail-tips may be absent (Rasmussen & Anderton 2005). Rufous cheeks and white supercilium are conspicuous at close quarters. Bill brown, and legs and feet are fleshy brown. Sexes are alike.

Distribution: It inhabits tall, damp grasslands of the Terai region and floodplains of large rivers, from western Nepal (Sukla Phanta), across the north Indian Terai to Assam and Manipur. It is also found in southeast China. In **Uttar Pradesh**, it is quite common in suitable habitats (e.g., **Katerniaghat**, **Dudhwa**) and may be common elsewhere. A detailed survey of the Terai is required.

Ecology: In an exhaustive study on the grassland birds of Nepal, Baral *et al.* (2006) encountered it in all types of relatively undisturbed, dense, tall grasslands with average sward height exceeding one metre. It was usually absent from moderately to heavily grazed grasslands. It also avoids homogeneous grasslands and shows preference for heterogeneous, complex grasslands. It is most frequent in *Phragmites-Saccharum* grasslands. Baral (2001) considers it an obligate grassland species as it was never found away from grasslands. The following brief description is based on Hem Sagar Baral's study in Nepal.

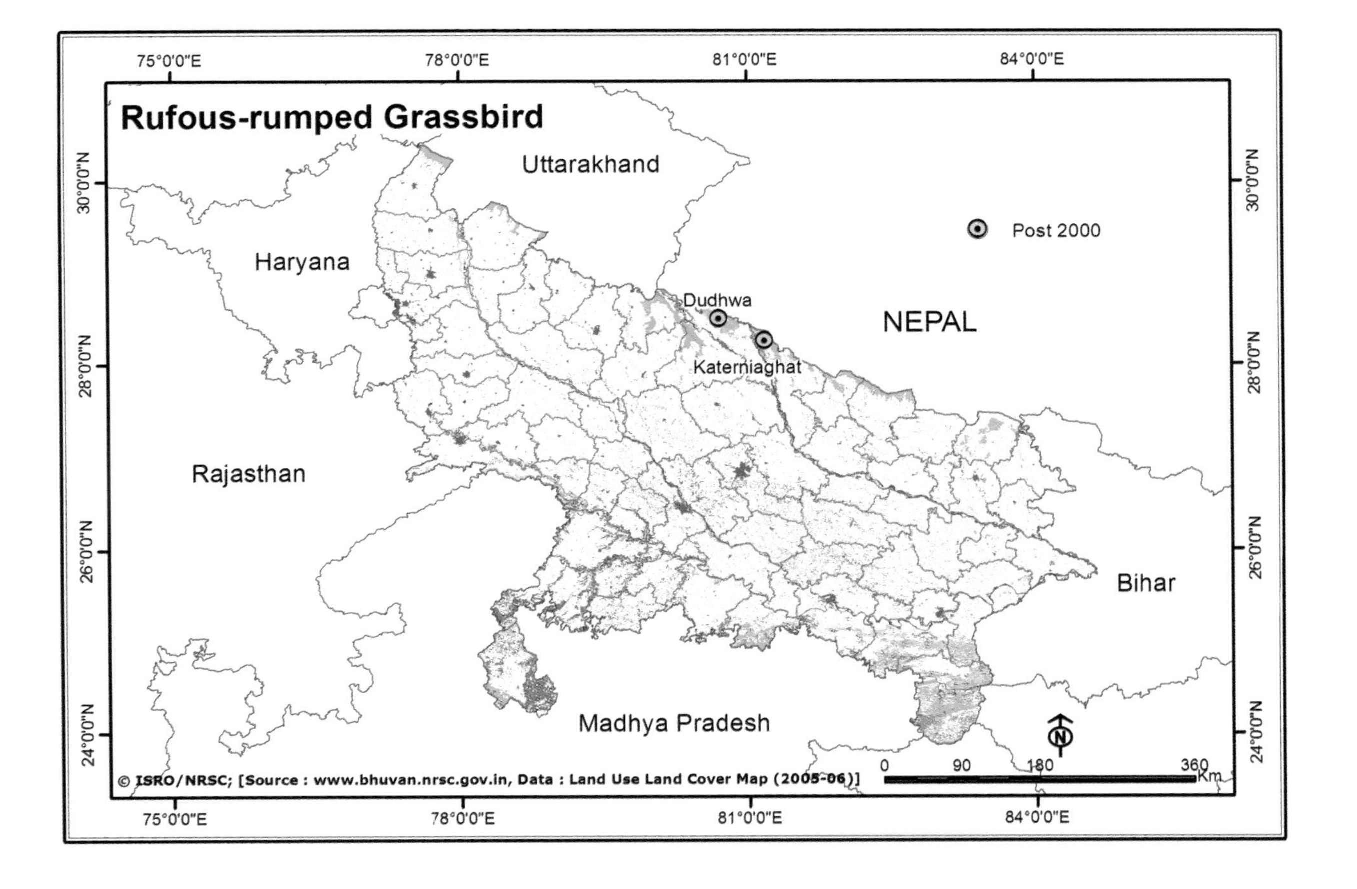

Rufous-rumped Grassbird
Uttarakhand
Haryana
Rajasthan
Madhya Pradesh
Bihar
NEPAL
Post 2000
Dudhwa
Katerniaghat
75°0'0"E
78°0'0"E
81°0'0"E
84°0'0"E
30°0'0"N
28°0'0"N
26°0'0"N
24°0'0"N
0
90
180
360
Km
© ISRO/NRSC; [Source : www.bhuvan.nrsc.gov.in, Data : Land Use Land Cover Map (2005-06)]

It lives singly or in pairs, and often utilises the lower half of the grass reeds and the open ground between clumps while feeding. Sometimes it is seen feeding on the ground in recently burnt and relatively open grasslands. During the breeding season, from April to August, mating display is usually seen in the morning and late afternoon when it is cool. Birds spread their wings, lower the head and body, fan out the tail, and often sing their melodious song at the same time. Normally, it remains hidden in tall grass, but in the breeding season its rare short display flight is undertaken about 3 m over the top of the grass, though it is not as elaborate as the flight of the Striated Grassbird *Megalurus palustris* or Bristled Grassbird *Chaetornis striata* (Baral 1997).

Threats: It is restricted to undisturbed grassland and is therefore susceptible to modification of its grassland and wetland habitat. Drainage, burning, overgrazing, and conversion to agriculture are the most pertinent threats (BirdLife International 2013). In all the areas where it is still found (Chitwan, Dudhwa, Kaziranga, Manas), the grassland is mainly maintained for large mammals such as Rhino, Swamp Deer, and Wild Buffalo, and sometimes burnt during the breeding season of the grassbird (starting in April), which could have a huge impact on its breeding success. Accidental catches may occur when trappers go for munias, pipits, or wagtails, which sometimes roost in tall grasslands, but this needs careful examination.

Conservation measures underway: Like all other wild birds, it is protected under the Indian Wildlife (Protection) Act, 1972. It is listed in Schedule IV of the Act. It is found in many PAs/IBAs in India and Nepal.

RECOMMENDATIONS

The following are specific recommendations for Uttar Pradesh:

(1) Conduct surveys in the whole Indian Terai, starting from Pilibhit in the west to Sohagi Barwa in the east.

(2) Identify key sites for intensive research on its ecology, behaviour, and microhabitat requirements, preferably with colour-marked birds and call playback.

(3) Study the impact of grassland burning regimes in controlled and experimental grassland plots. The extension areas of Dudhwa NP would be ideal for this study, so necessary management interventions can be taken up to make the habitat more suitable for this species.

(4) Initiate community-based protection of grasslands in multiple-use areas, particularly during summer when the bird breeds.

References

Ahmed, A. (1997) *Live Bird Trade in Northern India*. TRAFFIC India/WWF-India, New Delhi. Pp. 104.

Ahmed, A. (1999) *Fraudulence in Indian live bird trade: an identification monograph for control of illegal trade*. TRAFFIC India/WWF-India.

Ahmed, A. (2000) Illegal bird trade in India. *Mistnet* January–March 2000: 1–6.

Ahmed, A. (2002) Live Bird Trade in India. Unpublished report. TRAFFIC India/WWF-India, New Delhi.

Ahmed, A. (2012) Trade in Threatened Birds in India. Pp. 40–72. In: Rahmani, A.R. *Threatened Birds of India – Their Conservation Requirements*. Indian Bird Conservation Network: Bombay Natural History Society, Royal Society for the Protection of Birds and BirdLife International. Oxford University Press. Pp. xvi + 864.

Ahmed, A. and Rahmani, A.R. (1996) Illegal Trade of Raptors in India. Pp. 99–103. In: Rahmani, A.R. and Ugra, Gayatri (eds) *Birds of Wetlands and Grasslands: Proceedings of the Sálim Ali Centenary Seminar on Conservation of Avifauna of Wetlands and Grasslands*. Pp. x + 228. BNHS, Mumbai.

Ahmed, A., Rahmani, A.R. and Misra, M.K. (1996) Asian Red Data Birds in Indian Live Bird Trade. Paper presented at Pan-Asian Ornithological Conference and XII BirdLife Asia Conference, SACON, Coimbatore.

Ahmed, A., Rahmani, A.R., Das, G. and Misra, M.K. (1997) India's Illegal Falconry Trade. *TRAFFIC Bulletin* 17(1): 49–52.

Ali, S. (1935) Mainly in quest of Finn's Baya. *Indian Forester*: 365–374.

Ali, S. and Crook, J.H. (1959) Observations on Finn's Baya (*Ploceus megarhynchus* Hume) rediscovered in the Kumaon terai, 1959. *J. Bombay Nat. Hist. Soc.* 56: 457–483.

Ali, S. and Ripley, S.D. (1968) *Handbook of the Birds of India and Pakistan together with those of Nepal, Sikkim, Bhutan and Ceylon*. Divers to Hawks. (Sponsored by the Bombay Natural History Society) Oxford University Press, Bombay.

Ali, S. and Ripley S.D. (1974) *Handbook of the Birds of India and Pakistan*. Vol. 10. Oxford University Press, Bombay.

Ali, S. and Ripley, S.D. (1987) *Compact Handbook of the Birds of India and Pakistan together with those of Bangladesh, Nepal, Bhutan and Sri Lanka*. 2nd edn. Oxford University Press, Delhi. Pp. 890.

Ambedkar, V.C. (1968) Observations on the breeding biology of Finn's Baya (*Ploceus megarhynchus* Hume) in the Kumaon terai. *J. Bombay Nat. Hist. Soc.* 65: 596–607.

Anderson, A. (1875) on the occurrence of *Querquedula angustirostris* in the Doab and Oudh. *Stray Feathers* 3: 273.

Anon. (2004) Report on the International South Asian Vulture Recovery Plan Workshop. *Buceros* 9(1): 48.

Archibald, G.W., Sundar, K.S.G. and Barzen, J. (2003) A review of the three subspecies of Sarus Cranes *Grus antigone*. *Journal of Ecological Society* 16: 5–15.

Baker, E.C.S. (1921) *Game Birds of India, Burmah and Ceylon. Vol. 1. Ducks and their allies*. Bombay Natural History Society, Bombay.

Baker, E.C.S. (1924) *The Fauna of British India, including Ceylon and Burma. Birds*. Vol. 2. 2nd edn. Taylor and Francis, London.

Baker, E.C.S. (1928) *The Fauna of British India, including Ceylon and Burma. Birds*. Taylor and Francis, London.

Baker, E.C.S. (1929) *The Fauna of British India, including Ceylon and Burma. Birds.* Vol. 6. 2nd edn. Taylor and Francis, London.

Baral, H.S. (1997) Bristled Grassbirds *Chaetornis striatus* in Nepal. *Danphe* 6(2): 5–6.

Baral, H.S. (1998a) Finn's Weaver *Ploceus megarhynchus* and Singing Bushlark *Mirafra cantillans*: two new species for Nepal. *Forktail* 13: 129–131. Published 1998.

Baral, H.S. (1998b) Status, distribution, and habitat preferences of Swamp Francolin *Francolinus gularis* in Nepal. *Ibisbill* 1: 35–70.

Baral, H.S. (1999) Decline of wetland dependent birds in Nepal with reference to Chitwan. *Danphe* 8(1): 4–5.

Baral, H.S. (2001) Community structure and habitat associations of lowland grassland birds in Nepal. Universiteit van Amsterdam.

Baral, H.S. and Upadhyay, G.P. (2006) *Birds of Chitwan*. Department of National Parks and Wildlife Conservation and Bird Conservation Nepal, Kathmandu.

Baral, H.S., Wattel, J., Brewin, P. and Ormerod, S.J. (2006) Status, Distribution, Ecology and Behaviour of Rufous-rumped Grass Warbler *Graminicola bengalensis* Jerdon with reference to Nepal. *J. Bombay Nat. Hist. Soc.* 103(1): 44–48.

Baral, N., Gautam, R. and Tamang, B. (2005) Population status and breeding ecology of White-rumped Vulture *Gyps bengalensis* in Rampur Valley, Nepal. *Forktail* 21: 87–91.

Bates, R.S.P. and Lowther, E.H.N. (1952) *Breeding Birds of Kashmir*. Oxford University Press, Delhi.

Bhargava, R. (2000) A preliminary survey of the western population of Finn's Weaver in Kumaon terai, Uttar Pradesh, northern India. *OBC Bulletin* 32: 21–29.

Bhargava, R. (2004*) Assessing the threats and current status of Finn's Weaver* Ploceus megarhynchus *in India*. Indian Bird Conservation Network, Bombay Natural History Society, WWF-India and Birdlife International.

Bhargava, R. (2012) Live Bird Trade in India. Ph.D. Thesis submitted to the University of Mumbai, Mumbai. Pp. 508.

Bhatt, K. (2006) Notes on the nesting of Black-necked Stork (*Ephippiorhynchus asiaticus*) near the Marine National Park, Jodiya range, Jamnagar. *Newsletter for Birdwatchers* 46(2): 29–30.

BirdLife International (2001) *Threatened Birds of Asia: the Birdlife International Red Data Book*. Collar, N.J., Andreev, A.V., Chan, S., Crosby, M.J., Subramanya, S. and Tobias, J.A. (eds) Birdlife International, Cambridge, UK.

BirdLife International (2013) Species factsheets. Downloaded from http://www.birdlife.org.

BirdLife International (2014) IUCN Red List for birds. Downloaded from http://www.birdlife.org on 16/03/2014.

Bose, A.K., Curson, J. and Jarman, N. (1989) Report on birds in some national parks and other areas of special interest in India and Nepal '88–'89. Unpublished report.

Chauhan, R. and Andrews, H. (2006) Black-necked Stork *Ephippiorhynchus asiaticus* and Sarus Crane *Grus antigone* depredating eggs of the Three-striped Roofed Turtle *Kachuga dhongoka*. *Forktail* 22: 174–175.

Choudhury, A.U. (2000) *Birds of Assam*. Gibbon Books and WWF-India. Guwahati. Pp. 240.

Cunningham, A.H. (1928) Notes on duck shooting in the Roorkee district, U.P., in the years 1903 to 1927. *J. Bombay Nat. Hist. Soc.* 32: 600–605.

Cuthbert, R., Green, R.E., Ranade, S., Saravanan, S., Pain, D.J., Prakash, V. and Cunningham, A.A. (2006) Rapid population declines of Egyptian Vulture (*Neophron percnopterus*) and Red-headed Vulture (*Sarcogyps calvus*) in India. *Animal Conservation* 9: 349–354.

Cuthbert, R., Taggart, M.A., Prakash, V., Saini, M., Swarup, D., Upreti, S., Mateo, R., Chakraborty, S.S., Deori, P. and Green, R.E. (2011) Effectiveness of Action in India to Reduce Exposure of Gyps Vultures to the Toxic Veterinary Drug Diclofenac. PLoS ONE 6(5): e19069. doi:10.1371/journal.pone.0019069.

Datta, A. (1998) Hornbill abundance in unlogged forest, selectively logged forest and a forest plantation in Arunachal Pradesh, India. *Oryx* 32: 285–294.

Davidson, P. (2004) The distribution, ecology and conservation status of the Bengal Florican *Houbaropsis bengalensis* in Cambodia. M.Sc. Thesis. School of Environmental Sciences, University of East Anglia, Norwich, UK.

del Hoyo, J., Elliott, A. and Sargatal, J. (eds) (1992) *Handbook of Birds of the World.* Vol. 1: Ostriches to Ducks. Lynx Edicions, Barcelona.

del Hoyo, J., Elliott, A. and Sargatal, J. (eds) (1994) *Handbook of the Birds of the World*. Vol. 2: New World Vultures to Guineafowl. Lynx Edicions, Barcelona.

del Hoyo, J., Flliott A. and Sargatal J. (eds) (1996) *Handbook of the Birds of the World*. Vol. 3: Hoatzin to Auks. Lynx Edicions, Barcelona.

del Hoyo, J., Elliott, A. and Christie, D. (eds) (2005) *Handbook of the Birds of the World*. Vol. 10: Cuckoo-shrikes to Thrushes. Lynx Edicions, Barcelona.

Desai, J. H., Menon, G.H. and Shah, R. V. (1977) Studies on the reproductive pattern of the Painted Stork, *Ibis leucocephalus* (Pennant). *Pavo* 15(1&2): 1–32.

Donahue, J.P. (1967) Notes on a collection of Indian birds, mostly from Delhi. *J. Bombay Nat. Hist. Soc.* 64: 410–429.

Dutta, S., Rahmani, A., Gautam, P., Kasambe, R., Narwade, S., Narayan, G. and Jhala, Y. (2013) Guidelines for Preparation of State Action Plan for Resident Bustards Recovery Programme. Ministry of Environment and Forests, Government of India, New Delhi. Pp. 57+XI.

Ferguson-Lees, J. & Christie, D.A. (2001) *Raptors of the World*. Christopher Helm, London.

Ganguli, U. (1975) *A Guide to the Birds of the Delhi Area*. Indian Council of Agricultural Research, New Delhi.

Gilbert, M., Watson, R.T., Virani, M.Z., Oaks, J.L., Ahmed, S., Chaudhry, M.J.I., Arshad, M., Mahmood, S., Ali, A. and Khan, A.A. (2006) Rapid population declines and mortality clusters in three Oriental White-backed Vulture *Gyps bengalensis* colonies in Pakistan due to diclofenac poisoning. *Oryx* 40: 388–399.

Giri, T.R. and Choudhary, H. (1996) Additional sightings! *Bird Conservation Nepal Newsletter*. 5(3): 2–3.

Green, R.E., Newton, I., Shultz, S., Cunningham, A.A., Gilbert, M., Pain, D.J. and Prakash, V. (2004) Diclofenac poisoning as a cause of vulture population declines across the Indian subcontinent. *J. Appl. Ecol*. 41: 793–800.

Grewal, B. (1996) Bristled Grassbird *Chaetornis striatus* at Okhla, Delhi. *OBC Bulletin* 24: 43–44.

Grimmett, R., Inskipp, T. and Inskipp, C. (1999) *Birds of the Indian Subcontinent*. Oxford University Press, New Delhi. Pp. 384.

Grimmett, R., Inskipp, C. and Inskipp, T. (2011) *Birds of the Indian Subcontinent*. Oxford University Press, London. Pp. 528.

Hancock, J.A., Kushlan, J.A. and Kahl, M.P. (1992) *Storks, Ibises and Spoonbills of the World.* Academic Press, London.

Harris, C. (2001) Checklist of the birds of Yamuna river (Okhla to Jaitpur village). http://www.delhibirds.org/checklists/checklists_yamuna.htm.

Ilyas, O. and Khan, J.A. (2006) Birds of Chandraprabha Wildlife Sanctuary. *Newsletter for Birdwatchers* 46(4): 51–52.

Inskipp, C. and Inskipp, T. (1983) Report on a Survey of Bengal Florican (*Houbaropsis bengalensis*) in Nepal and India, 1982. ICBP Study Report No. 2. Pp. 54.

Inskipp, C. and Inskipp, T. (1991) *A Guide to the Birds of Nepal*. 2nd edn. Christopher Helm and A&C Black, London. Pp. 400.

Iqubal, Perwez, McGowan, Philip J.K., Carroll, John P. and Rahmani, Asad R. (2003) Home range size, habitat use and nesting success of Swamp Francolin *Francolinus gularis* on agricultural land in northern India. *Bird Conservation International* 13: 127–138.

Ishtiaq, F. (1998) Comparative ecology and behaviour of storks in Keoladeo National Park, Rajasthan, India. Ph.D. Thesis. Aligarh Muslim University, Aligarh, India.

Ishtiaq, F. (2009) Avian malaria and decline of White-backed Vulture population. *Curr. Sci.* 97(2): 134–135.

Ishtiaq, F., Javed, S., Coulter, M.C. and Rahmani A.R. (2010) Resource partitioning in three sympatric species of storks in Keoladeo National Park, India. *Waterbirds* 33(1): 41–49.

Ishtiaq, F., Rahmani, A.R., Javed, S. and Coulter, M.C. (2004) Nest-site characteristics of Black-necked Stork *Ephippiorhynchus asiaticus* and Woolly-necked Stork *Ciconia episcopus* in Keoladeo National Park, Bharatpur, India. *J. Bombay Nat. His. Soc.* 101(1): 90–95.

Islam, M.Z. and Rahmani, A.R. (2004) *Important Bird Areas in India: Priority sites for conservation.* Indian Bird Conservation Network, BNHS, and BirdLife International, UK. Pp. 1133.

Islam, M.Z. and Rahmani, A.R. (2008) *Potential and existing Ramsar sites in India*. Indian Bird Conservation Network: Bombay Natural History Society, BirdLife International and Royal Society for the Protection of Birds. Oxford University Press, Bombay Pp. 592

Islam, M.Z., Ugra, G., Rahmani, A.R. and Prakash, V. (1999) *An Illustrated Guide to the Birds of Mathura Refinery*. Bombay Natural History Society, Mumbai.

Javed, S. (1996) The Swamp Francolin – a bird to watch. *Sanctuary Asia* 16(4): 57–60.

Javed, S. (2000) Current status and distribution of Swamp Francolin (*Francolinus gularis*) in the North Indian terai. Department of Wildlife Sciences, Aligarh Muslim University, Aligarh.

Javed, S. (2001). Eye in the sky. *Sanctuary Asia* XXI (1): 36–40.

Javed, S. and Rahmani, A.R. (1991) Swamp Francolin in the north Indian terai. *World Pheasant Assoc. News* 34: 15–18.

Javed, S. and Rahmani, A.R. (1998) Conservation of the avifauna of Dudwa National Park, India. *Forktail* 14: 55–64.

Javed, S., Qureshi, Q. and Rahmani, A.R. (1999) Conservation status and distribution of Swamp Francolin in India. *J. Bombay Nat. Hist. Soc.* 96: 16–23.

Johnsgard, P.A. (1991) *Bustards, Hemipodes, and Sandgrouse: Birds of Dry Places*. Oxford University Press, Oxford.

Johnson J.A., Lerner, H.R.L., Rasmussen, P.C. and Mindell, D.P. (2006) Systematics within *Gyps* vultures: a clade at risk. *BMC Evolutionary Biology* 6: 65.

Kalam, A. and Urfi, A.J. (2008) Foraging behaviour and prey size of the Painted Stork. *Journal of Zoology* 274(2): 198–204.

Kaur, J. (2008) Impact of Land Use Changes on the Habitat, Behaviour and Breeding Biology of the Indian Sarus Crane (*Grus antigone antigone*) in the Semi-arid Tract of Rajasthan, India. Ph.D. Thesis. Forest Research Institute University, Dehra Dun.

Kaur, J. and Choudhury, B.C. (2003) Recognition of community involvement in Sarus Crane conservation in Kota, Rajasthan. *Mistnet* 3(4): 6.

Kaur, J. and Choudhury, B.C. (2005) Predation by Marsh Harrier *Circus aeruginosus* on chick of Sarus Crane *Grus antigone* in Kota, Rajasthan. *J. Bombay Nat. His. Soc.* 102(1): 102–103.

Kaur, J. and Nair, A. (2008) Community involvement in conservation of Sarus Crane breeding habitat in three districts of semi-arid tract of Rajasthan, India. Report submitted to Rufford Small Grants Foundation, UK.

Kaur, J., Nair, A. and Choudhury, B.C. (2008) Conservation of the vulnerable Sarus Crane *Grus antigone antigone* in Kota, Rajasthan, India: a case study of community involvement. *Oryx* 42: 452–455.

Kemp, A.C. (1995) *The Hornbills*. Series: Bird Families of the World, Vol. 1. Oxford University Press, Oxford.

Khan, S.A. (1992) Ecological Studies on *Grus antigone* Sarus Crane with Special Reference to His (sic) Breeding Behaviour. Pp. 152. Ph.D. Thesis. Y.D. College, Lakhimpur Kheri, Uttar Pradesh.

Kotagama, S. and Ratnavira, G. (2010) *An Illustrated Guide to the Birds of Sri Lanka*. Field Ornithology Group of Sri Lanka, Colombo, Sri Lanka.

Kumar, R., Shahabuddin, G. and Kumar, A. (2011) How good are managed forests at conserving native woodpecker communities? A study in Sub-Himalyan dipterocarp forests of northwest India. *Biological Conservation* 144: 1876–1884.

Kumar, S. and Srivastav, N. (2011) *Conservation of Potential wetlands in district Sitapur*. U.P. State Biodiversity Board. Pp. 143.

Lahkar, B.P. (2000) Pallas's Fishing Eagle *Haliaeetus leucoryphus* (Pallas) pirates fish from an Otter *Lutra lutra* (Linn.). *J. Bombay Nat. His. Soc.* 97(3): 425.

Lammertink, M. (2004) A multiple-site comparison of woodpecker communities in Bornean lowland and hill forests. *Conserv. Biol.* 18: 746–757.

Lethaby, N. (2005) The occurrence of Lesser Fish Eagle *Ichthyophaga humilis* on the Cauvery river, Karnataka, India and some notes on the identification of this species. *BirdingASIA* 4: 33–38.

Maheswaran, G. (1998) Ecology and behaviour of the Black-necked Stork *Ephippiorhynchus asiaticus* in Dudhwa National Park, Uttar Pradesh, India. Ph.D. Thesis. Aligarh Muslim University, Aligarh, India.

Maheswaran, G. and Rahmani, A.R. (2001) Effects of water level changes and wading bird abundance on the foraging behaviour of blacknecked storks *Ephippiorhynchus asiaticus* in Dudhwa National Park, India. *Journal of Biosciences* (Bangalore). 26(3): 373–382.

Manakadan, R. and Kannan, V. (2003) A study of Spot-billed Pelican *Pelecanus philippensis* with special reference to its conservation in southern India. Final Report. BNHS, Mumbai.

Manu, K. and Jolly, S. (2000). Pelicans and People: The two-tier village of Kokkare-Bellur, Karnataka, India. Community based conservation in south Asia: Case Study No. 4. Kalpavriksh and International Institute of Environment and Development.

McGowan, P.J.K. and Garson, P.J. (1995) *Pheasants: status survey and conservation action plan 1995–1999.* IUCN–The World Conservation Union, Gland, Switzerland.

McGowan, P.J.K., Javed, S. and Rahmani, A.R. (1995) Swamp Francolin *Francolinus gularis* survey technique: a case study from Northern india. *Forktail* 11: 101–110. Published 1996.

Meganathan, T. and Urfi, A.J. (2009) Inter-colony variations in nesting ecology of Painted Stork (*Mycteria leucocephala*) in the Delhi zoo (north India) *Waterbirds* 32(2): 352–356.

Meine, C.D. and Archibald, G.W. (eds) (1996) *The Cranes: status survey and conservation action plan.* IUCN–The World Conservation Union, Gland, Switzerland, and Cambridge, UK.

Mukherjee, A., Borad, C.K. and Parasharya, B.M. (2001) The Indian Sarus Crane (*Grus antigone antigone*) – a threatened wetland species. *Newsletter for Birdwatchers* 40(6): 77–78.

Muralidharan, S. (1992) Poisoning the Sarus. *Hornbill* 1992(1): 2–7.

Muralidharan, S., Dhananjayan, V., Riseborough, R., Prakash, V. Jayakumar, R. and Bloom, P.H. (2008) Persistent organochlorine pesticide residues in tissues and eggs of White-backed Vultures *Gyps bengalensis* from different locations in India. *Bull. Environ. Contam. Toxicol.* 81: 561–565.

Nagulu, V. (1983) Feeding and breeding biology of Grey Pelican at Nelapattu Bird Sanctuary in Andhra Pradesh, India. Ph.D. Thesis. Osmania University, Hyderabad.

Naoroji, R. (1997) Contamination in egg shells of Himalayan Greyheaded Fishing Eagle *Ichthyaetus nana plumbea* in Corbett National Park, India. *J. Bombay Nat. Hist. Soc.* 94: 398–400.

Naoroji, R. (2007) *Birds of Prey of the Indian Subcontinent*. Om Books International, New Delhi.

Narayan, G. (1992) Ecology, distribution and conservation of the Bengal Florican *Houbaropsis bengalensis* (Gmelin) in India. Ph.D. Thesis. University of Bombay, Bombay.

Narayan, G. and Rosalind, L. (1990) The Bengal Florican at Manas Wildlife Sanctuary. Pp. 35–43. In: Status and ecology of the Lesser and Bengal Floricans with reports on Jerdon's Courser and Mountain Quail. Final Report. Bombay Natural History Society, Bombay.

Narayan, G. and Rosalind, L. (1997) Wintering range and time extension of Hodgson's Bush Chat *Saxicola insignis* Gray in India. *J. Bombay Nat. Hist. Soc.* 94: 572–573.

Neelakantan, K.K. (1949) A south Indian pelicanry. *J. Bombay Nat. Hist. Soc.* 48: 656–666.

Oaks, J.L., Gilbert, M., Virani, M.Z., Watson, R.T., Meteyer, C.U., Rideout, B., Shivaprasad, H.L., Ahmed, S., Chaudhry, M.J.I., Arshad, M., Mahmood, S., Ali, A. and Khan, A.A. (2004a) Diclofenac residues as the cause of vulture population decline in Pakistan. *Nature* 427: 630–633.

Oaks, J.L., Donahue, S.L., Rurangirwa, F.R., Rideout, B.A., Gilbert, M. and Virani, M.Z. (2004b) Identification of a novel mycoplasma species from an Oriental Whitebacked Vulture (*Gyps bengalensis*). *J. Clinical Microbiol.* 42: 5909–5912.

Pain, D.J., Cunningham, A.A., Donald, P.F., Duckworth, J.W., Houston, D.C., Katzner, T., Parry-Jones, J., Poole, C., Prakash, V., Round, P. and Timmins, R. (2004) Causes and Effects of Temporospatial Declines of *Gyps* Vultures in Asia. *Conservation Biology* 17(3): 661–671.

Pain, D.J., Bowden, C.G.R., Cunningham, A.A., Cuthbert, R., Das, D., Gilbert, M., Jakati, R.D., Jhala, Y., Khan, A.A., Naidoo, V., Oaks, J.L., Parry-Jones, J., Prakash, V., Rahmani, A., Ranade, S.P., Baral, H.S., Senacha, K.R. and Saravanan, S. (2008) The race to prevent the extinction of South Asian vultures. *Bird Conservation International* 18: 30–48.

Parry, S.J., Clark, W.S. and Prakash, V. (2002) On the taxonomic status of the Indian Spotted Eagle *Aquila hastata*. *Ibis* 144(4): 665–675.

Pasha, M.K.S. (1995) A preliminary avifaunal survey in and around Bijnor, Uttar Pradesh. *Newsletter for Birdwatchers* 35: 25–28.

Peet, N.B. (1997) Biodiversity and management of tall grasslands in Nepal. Ph.D. Thesis. University of East Anglia, Norwich, UK.

Poharkar, A., Reddy, P.A., Gadge, V.A., Kolte, S., Kurkure, N. and Shivaji, S. (2009) Is malaria the cause for decline in the wild population of the Indian White-backed Vulture (*Gyps bengalensis*) *Curr. Sci.* 96: 553–558.

Prakash, V. (1989) Population and distribution of raptors in Keoladeo National Park, Bharatpur, India. Pp. 129–137. In: Meyburg, B.U. and Chancellor, R.D. (eds) *Raptors in the Modern World. Proceedings of the III World Conference on Birds of Prey and Owls*, Eilat, Israel, 22–27 March, 1987. World Working Group on Birds of Prey, Berlin.

Prakash, V. (1999) Status of vultures in Keoladeo National Park, Bharatpur, Rajasthan, with special reference to population crash in Gyps species *J. Bombay Nat. His. Soc.* 96 (3): 364–365.

Prakash, V., Green, R.E., Pain, D.J., Ranade, S.P., Saravanan, S. and Prakash, N. (2007) Recent changes in populations of resident *Gyps* vultures in India. *J. Bombay Nat. His. Soc.* 104: 127–133.

Prakash, V., Pain, D.J., Cunningham, A.A., Donald, P.F., Prakash, N., Verma, A., Gargi, R., Sivakumar, S. and Rahmani, A.R. (2003) Catastrophic collapse of Indian White-backed *Gyps bengalensis* and Long-billed *Gyps indicus* Vulture populations. *Biological Conservation* 109(3): 381–390.

Rahmani, A.R. (1989) Status of the Black-necked Stork *Ephippiorhynchus asiaticus* in the Indian subcontinent. *Forktail* 5: 99–110.

Rahmani, A.R. (1992) *The Wetlands of Uttar Pradesh* – Part III. NLBW 32(1): 3–5.

Rahmani, A.R. (2008) Race to save Vultures. *Hornbill* (Oct-Dec) 2008: 147–155.

Rahmani, A.R. (2012) *Threatened Birds of India – Their Conservation Requirements*. Indian Bird Conservation Network: Bombay Natural History Society, Royal Society for the Protection of Birds and BirdLife International. Oxford University Press. Pp. xvi+864.

Rahmani, A.R. and Islam, M.Z. (2008) *Ducks, Geese and Swans of India*. Indian Bird Conservation Network, Bombay Natural History Society, Royal Society for the Protection of Birds, and BirdLife International. Oxford University Press, Delhi. Pp. 374.

Rahmani, A.R., Kumar, S., Deori, P., Khan, J.A., Kalra, M., Belal, M.S., Khan, A.M., Khan, N.I., George, A., Srivastava, N., Singh, V.P., Rehman, F. and Muralidharan, S. (2010) *Migratory Movements of Waterbirds through Uttar Pradesh and the Surveillance of Avian Diseases*. Bombay Natural History Society, Mumbai. Pp. 405.

Rahmani, A.R., Narayan, G., Rosalind, L. and Sankaran, R. (1990) Status of the Bengal Florican in India. Pp. 55-78. In: Status and Ecology of the Lesser and Bengal Floricans. Final Report. Bombay Natural History Society, Bombay.

Rai, Y.M. (1979) Observations on Finn's Baya breeding near Meerut. *Newsletter for Birdwatchers* 19(11): 11.

Rai, Y.M. (1983) Breeding notes (the fragile nature balance). *Newsletter for Birdwatchers* 23(9-10): 17.

Ramachandran, N.K. and Vijayan, V.S. (1994) General ecology of the Sarus Crane at Keoladeo National Park, Bharatpur, India. In: *Proc. International Crane Workshop, China.* May 1987. (ed.) International Crane Foundation, Wisconsin, USA.

Rana, G. and Prakash V. (2004) Unusually high mortality of Cranes in areas adjoining Keoladeo National Park, Bharatpur, Rajasthan. *J. Bombay Nat. His. Soc.* 101 (2): 317.

Rasmussen, P.C. and Anderton, J.C. (2005) *Bird of South Asia: The Ripley Guide*. 2 vols. Smithsonian Institution, Washington DC and Lynx Edicions, Barcelona.

Rasmussen, P.C. and Anderton J.C. (2012) *Birds of South Asia: The Ripley Guide* Vol. 1: Field guide. Vol. 2: Attributes and status. Smithsonian Institution, Washington DC, Michigan State University, Michigan and Lynx Edicions, Barcelona.

Reid, G. (1881) The birds of the Lucknow Civil Division. *Stray Feathers* 9: 491–504; 10: 1–88.

Roberts, T.J. (1991) A supplementary note on Khar, Pakistan. *OBC Bull.* 14: 35–37.

Sankaran, R. (1991) Some aspects of the breeding behaviour of the Lesser Florican *Sypheotides indica* (J.F. Miller) and the Bengal Florican *Eupodotis bengalensis* (Gmelin). Ph.D. Thesis. University of Bombay, Bombay.

Sankaran, R. (1996) Territorial displays of the Bengal Florican. *J. Bombay Nat. Hist. Soc.* 93: 167–177.

Sankaran, R. and Rahmani, A.R. (1990) The Bengal Florican in Dudwa National Park. Pp. 45–54. In: Status and ecology of the Lesser and Bengal Floricans with reports on Jerdon's Courser and Mountain Quail. Final Report. Bombay Natural History Society, Bombay.

Sankaran, R., Rahmani, A.R. and Ganguli-Lachungpa, U. (1992) The distribution and status of the Lesser Florican *Sypheotides indica* (J.F. Miller) in the Indian subcontinent. *J. Bombay Nat. Hist. Soc.* 89: 156–179.

Seibold, I. and Helbig, A.J. (1995) Evolutionary history of New and Old World vultures inferred from nucleotide sequences of the Mitochondrial Cytochrome b Gene. *Philosophical Transactions: Biological Sciences* 350 (1332): 163–178.

Senacha, K.R., Taggart, M.A., Rahmani, A.R., Jhala, Y.V., Cuthbert, R., Pain, D.J. and Green, R.E. (2008) Diclofenac levels in livestock carcasses in India before the 2006 "ban". *J. Bombay Nat. Hist. Soc.* 105(2): 148–161.

Shahabuddin, G. and Kumar, R. (2011) Assessing conservation threat in an Endemic Bird Area: The Great Slaty Woodpecker in Sub-Himalyan Uttarakhand, India. Final Report for WWF-India. Pp. 18.

Sharma, A.K. (1984) Migratory avifauna of Meerut, India. *Tigerpaper* 11(4): 15–17.

Sharma, M. (2007) Bristled Grassbird *Chaetornis striatus* in Corbett National Park, India. *BirdingAsia* 7: 90–91.

Sharma, R.K. and Singh, L.A.K. (1986) Wetland birds in National Chambal Sanctuary. Preliminary report. Unpublished.

Sharma, R.K., Sharma, S. and Mathur, R. (1995) Faunistic survey of river Mahanadi vis-à-vis environmental conditions in Madhya Pradesh. *Tigerpaper* 22(3): 21–26.

Shivprakash, A., Das, K.R.K., Shivanand, T., Girija, T. and Sharath, A. (2006) Notes on the breeding of the Indian Spotted Eagle *Aquila hastata*. *Indian Birds* 2(1): 2–4.

Shultz, S., Baral, H.S., Charman, S., Cunningham, A.A. , Das, D., Ghalsasi, D.R., Goudar, M.S., Green, R.E., Jones, A., Nighot, P., Pain, D.J. and Prakash, V. (2004) Diclofenac poisoning is widespread in declining vulture populations across the Indian subcontinent. *Proceedings of the Royal Society of London*, B (Suppl.), 271: 458–460.

Subramanya, S. (1996a) Heronries of Andhra Pradesh. *Mayura* 13: 1–27.

Subramanya, S. (1996b) Distribution, status and conservation of Indian heronries. *J. Bombay Nat. Hist. Soc.* 93: 459–486.

Sundar, G.V. and Pandav, B. (2007) Observations on breeding biology of three stork species in Bhitarkanika mangroves. *Indian Birds* 3(2): 45–50.

Sundar, K.S.G. (1999) Black-necked Storks, Sarus Cranes and Drongo Cuckoos. *Newsletter for Birdwatchers* 39(5): 71–72.

Sundar, K.S.G. (2003) Notes on the breeding biology of the Black-necked Stork *Ephippiorhynchus asiaticus* in Etawah and Mainpuri districts, Uttar Pradesh, India. *Forktail* 19: 15–20.

Sundar, K.S.G. (2004) Observations on breeding Indian Skimmers *Rynchops albicollis* in the National Chambal Sanctuary, Uttar Pradesh, India. *Forktail* 20: 89–90.

Sundar, K.S.G. (2005a) Effectiveness of road transects and wetland visits for surveying Black-necked Stork *Ephippiorhynchus asiaticus* and Sarus Crane *Grus antigone* in India. *Forktail* 21: 27–32.

Sundar, K.S.G. (2005b) Predation of fledgling Painted Stork *Mycteria leucocephala* by a Spotted Eagle *Aquila* sp. in Sultanpur National Park, Haryana. *Indian Birds* 1(6): 144–145.

Sundar, K.S.G. (2006) The Lesser Florican *Sypheotides indica* in Mainpuri, Uttar Pradesh, India. *Indian Birds* 2(1): 10.

Sundar, K.S.G. (2009) Are rice paddies suboptimal breeding habitat for Sarus Cranes in Uttar Pradesh, India? *The Condor* 111: 611–623.

Sundar, K.S.G. and Choudhury, B.C. (2001) A note on Sarus Crane *Grus antigone* mortality due to collision with high-tension power lines. *J. Bombay Nat. Hist. Soc.* 98: 108–110.

Sundar, K.S.G. and Choudhury, B.C. (2003) The Indian Sarus Crane *Grus a. antigone*: a literature review. *J. Ecol. Soc.* 16: 16–41.

Sundar, K.S.G. and Choudhury, B.C. (2005) Effect of incubating adult sex and clutch size on egg orientation in Sarus Cranes *Grus antigone*. *Forktail* 21: 179–181.

Sundar, K.S.G. and Choudhury, B.C. (2006) Conservation of the Sarus Crane *Grus antigone* in Uttar Pradesh, India. *J. Bombay Nat. Hist. Soc.* 103(2&3): 182–190.

Sundar, K.S.G., Kaur, J. and Choudhury, B.C. (2000) Distribution, demography and conservation status of the Indian Sarus Crane (*Grus antigone antigone*) in India. *J. Bombay Nat. Hist. Soc.* 97(3): 319–339.

Sundar, K.S.G., Deomorari, A., Bhatia, Y. and Narayanan, S.P. (2007) Records of Black-necked Stork *Ephippiorhynchus asiaticus* breeding pairs fledging four chicks. *Forktail* 23: 161–163.

Swan, G., Naidoo, V., Cuthbert, R., Green, R.E., Pain, D.J., Swarup, D., Prakash, V., Taggart, M., Bekker, L., Das, D., Diekmann, J., Diekmann, M., Killian, E., Meharg, A., Patra, R.C., Saini, M. and Wolter, K. (2006a) Removing the threat of diclofenac to critically endangered Asian vultures. *PLoS Biol.* 4(3): e66.doi: 10,1371/journal. pbio.0040066.

Swan, G.E., Cuthbert, R., Quevedo, M., Green, R.E., Pain, D.J., Bartels, P., Cunningham, A.A., Duncan, N., Meharg, A.A., Oaks, J.L., Parry-Jones, J., Shultz, S., Taggart, M.A., Verdoorn, G. and Wolter, K. (2006b) Toxicity of diclofenac to *Gyps* vultures. *Biol. Lett.* 2(2): 279–282.

Taggart, M.A., Cuthbert, R., Das, D., Pain, D.J., Green, R.E., Shultz, S., Cunningham, A.A. and Meharg, A.A. (2006) DIclofenac disposition in Indian cow and goat with reference to *Gyps* vulture population declines. *Environ. Pollut.* 147: 6065.

Talukdar, B.K. (1995) Spot-billed Pelican in Assam. *OBC Bull.* 22: 46–47.

Talukdar, B.K. (1999): The status of Spot-billed Pelican in Assam, India. *OBC Bull.* 30: 13–14.

Urfi, A.J. (1996) On some new breeding records of waterbirds from the Delhi region. *J. Bombay Nat. Hist. Soc.* 93 (1): 94–95.

Urfi , A.J. (2003) The birds of Okhla barrage bird sanctuary, Delhi, India. *Forktail* 19: 39–50.

Urfi, A.J. and Kalam, A. (2006) Sexual size dimorphism and mating pattern in the Painted Stork (*Mycteria leucocephala*). *Waterbirds* 29: 489–496.

Urfi, A.J., Meganathan, T. and Kalam, A. (2007) Nesting ecology of the Painted Stork *Mycteria leucocephala* at Sultanpur National Park, Haryana, India. *Forktail* 23: 150–153.

Urquhart, E. (2002) *Stonechats: A Guide to the Genus* Saxicola. Christopher Helm, London. Pp. 320.

Vali, U. (2006) Mitochondrial DNA sequence support Species status for the Indian Spotted Eagle *Aquila hastata*. *Bull. British Ornithological Union* 126(3): 238–242.

■ ■ ■

Index of Common Names

Index of Scientific Names

ABOUT THE BOMBAY NATURAL HISTORY SOCIETY

The BNHS was founded in 1883 and today it is the prime non-governmental conservation organisation in the Subcontinent. We work towards the conservation of nature and natural resources, education and research in natural history, and have members in over 20 countries.

Membership Activities and Benefits

- Nature camps to wildlife places both in and outside India.
- Treks, walks and field trips at weekends.
- Excellent audio-visuals presented by experts regularly.
- Seminars, workshops and correspondence courses on specific natural history subjects.
- Members receive *Hornbill*, a quarterly magazine.
- Subscription to the *Journal* is optional to members.
- Up to 15% discount on BNHS publications.
- 10% discount on BNHS products.
- Access to the finest collection of books on natural history.
- Voluntary Nature Education and Conservation activities.

Publications

BNHS Publications have been the standard reference works on the natural history of the Indian subcontinent since 1886. They are essential acquisitions for naturalists, amateurs and professionals throughout the country and abroad. Published uninterrupted since 1886, the *Journal of the Bombay Natural History Society* is acknowledged to be one of the finest scientific natural history sources for the Oriental Region. The popular quarterly magazine *Hornbill*, published since 1976, caters to a varied readership of all ages.

***To become a member or for other details contact*:**

Bombay Natural History Society

Hornbill House, S.B. Singh Road, Mumbai 400 001, Maharashtra, India.

Tel.: +91-22-2282 1811 Fax: +91-22-2283 7615

Email: info@bnhs.org Website: www.bnhs.org

THE SOCIETY'S PUBLICATIONS

		List Price
1.	**The Book of Indian Birds** by Sálim Ali, 13th edition	Rs. 495
2.	**Birds of the Indian Subcontinent – A Field Guide** by Ranjit Manakadan, J.C. Daniel & Nikhil Bhopale	Rs. 550
3.	**A Guide to the Cranes of India** by Prakash Gole	Rs. 75
4.	**Birds of Wetlands and Grasslands** edited by Asad R. Rahmani & Gayatri Ugra	Rs. 500
5.	**Indian Bird Banding Manual** Compiled by S. Balachandran	Rs. 100
6.	**Birds of Sanjay Gandhi National Park** by Sunjoy Monga	Rs. 50
7.	**Important Bird Areas in India: Priority Sites for Conservation** edited by M. Zafar-ul Islam & Asad R. Rahmani	Rs. 3000
8.	**Important Bird Areas of Maharashtra: Priority Sites for Conservation** by Asad R. Rahmani, M. Zafar-ul Islam, Raju Kasambe & Jayant Wadatkar	Rs. 300
9.	**Potential and Existing Ramsar Sites in India** by M. Zafar-ul Islam & Asad R. Rahmani	Rs. 1500
10.	**Ducks, Geese and Swans of India** by M. Zafar-ul Islam & Asad R. Rahmani	Rs. 1300
11.	**Threatened Birds of India (Their Conservation Requirements)** by Asad R. Rahmani	Rs. 3000
12.	**Threatened Birds of Assam** by Asad R. Rahmani & Anwaruddin Choudhury	Rs. 200
13.	**Threatened Birds of Maharashtra** by Asad R. Rahmani, Raju Kasambe, Sujit Narwade, Pramod Patil & Noor I. Khan	Rs. 300
14.	**Threatened Birds of Uttarakhand** by Asad R. Rahmani & Dhananjai Mohan	Rs. 300
15.	**Bird Atlas of Chilka** (Paperback) by S. Balachandran & P. Sathiyaselvam	Rs. 250
16.	**Bird Atlas of Chilka** (Hardbound) by S. Balachandran & P. Sathiyaselvam	Rs. 300
17.	**Latin Names of Indian Birds – Explained** by Satish Pande	Rs. 325
18.	**Birds of the Great Andamanese – Names, Classification and Culture** by Satish Pande & Anvita Abbi	Rs. 950
19.	**Bharat ke Pakshi (Hindi)** Translated by Gayatri Ugra	Rs. 500
20.	**Uttar Bharat na Pakshiyo (Gujarati)** by Richard Grimmett, Tim Inskipp & Sarita Sharma	Rs. 500
21.	**Shumali Hindustan ke Parinde (Urdu)** by Richard Grimmett, Tim Inskipp & M. Zafar-ul Islam	Rs. 500

List Price

22. **Thekke Indiayile Pakshikal (Malayalam)**
by Richard Grimmett, Tim Inskipp & P.O. Nameer — Rs. 500

23. **Then Indiye Paravaigal (Tamil)**
by Richard Grimmett, Tim Inskipp & Gopinathan Maheswaran — Rs. 500

24. **Dakshin Bharat Pakshilu (Telugu)**
by Richard Grimmett, Tim Inskipp & Birdwatchers Society of Andhra Pradesh — Rs. 500

25. **Petronia**
edited by J.C. Daniel & Gayatri Ugra — Rs. 400

26. **Loke Wan Tho's Birds**
Photo editors: Rishad Naoroji & Hira Punjabi — Rs. 2000

27. **Important Bird Areas of Jammu & Kashmir (Priority Sites for Conservation)**
by Asad R. Rahmani, M. Zafar-ul Islam, Khursheed Ahmed, Intesar Suhail, Pankaj Chandan & Ashfaq Ahmed Zarri — Rs. 250

28. **The Book of Indian Animals**
by S.H. Prater, 3rd edition — Rs. 360

29. **A Week with Elephants — Proceedings of the Seminar on Asian Elephants, June 1993**
edited by J.C. Daniel & Hemant Datye — Rs. 450

30. **The Book of Indian Reptiles and Amphibians**
by J.C. Daniel — Rs. 595

31. **The Book of Indian Trees**
by K.C. Sahni, 2nd edition — Rs. 295

32. **Common Indian Wild Flowers**
by Isaac Kehimkar — Rs. 400

33. **The Trees of Mumbai**
by Marselin Almeida & Naresh Chaturvedi — Rs. 465

34. **Illustrated Flora of Keoladeo National Park, Bharatpur**
by V.P. Prasad, Daniel Mason, Joy E. Marburger & C.R. Ajithkumar — Rs. 695

35. **The Book of Indian Butterflies**
by Isaac Kehimkar — Rs. 1500

36. **Understanding the Sea**
by B.F. Chhapgar — Rs. 500

37. **Treasures of Indian Wildlife**
edited by A.S. Kothari & B.F. Chhapgar — Rs. 1900

38. **Living Jewels from the Indian Jungle**
edited by A.S. Kothari & B.F. Chhapgar — Rs. 1600

39. **Wildlife of the Himalayas and the Terai Region**
edited by A.S. Kothari & B.F. Chhapgar — Rs. 1250

40. **Natural History and the Indian Army**
by J.C. Daniel & Lt. Gen. Baljit Singh (ERetd.) — Rs. 1200

41. **Cassandra of Conservation**
edited by J.C. Daniel — Rs. 200

		List Price
42.	**National Parks and Sanctuaries in Maharashtra Vol. I & II** by Pratibha Pande	Rs. 500
43.	**In Harmony with Nature – A Teacher's Handbook** by BNHS Conservation Education Centre	Rs. 350
44.	**Maitri Nisargashi – A Teacher's Handbook (Marathi)** by BNHS Conservation Education Centre	Rs. 350
45.	**Green Guide for Teachers (English)** by CEC team, Goregaon	Rs. 350
46.	**Nisarg Margdarshak for Teachers (Marathi)** by CEC team, Goregaon	Rs. 350
47.	**100 Volumes of JBNHS (DVD)**	Rs. 1000
48.	**Indian Bird Calls (DVD)**	Rs. 500

Notes

Notes

Notes